Student's Solutions Guide for Introduction to Probability, Statistics, and Random Processes

Hossein Pishro-Nik
University of Massachusetts Amherst

Published by Kappa Research, LLC.

Printed in the United States of America

ISBN: 978-0-9906372-1-9

Contents

Preface

In this book, you will find guided solutions to the odd-numbered end-of-chapter problems found in the companion textbook, *Introduction to Probability, Statistics, and Random Processes.*

Since the textbook's initial publication in 2014, I have received many requests to publish the solutions to those problems. I have published this book so that students may learn at their own pace with guided help through many of the problems presented in the original text.

It is my hope that this book serves its purpose well and enables students to access help to these problems. To access the original textbook as well as video lectures and probability calculators please visit www.probabilitycourse.com.

Acknowledgements

I would like to thank Laura Handly and Linnea Duley for their detailed review and comments. I am thankful to all of my teaching assistants who helped in various aspects of both the course and the book.

Chapter 1

Basic Concepts

1. Suppose that the universal set S is defined as $S = \{1, 2, \cdots, 10\}$ and $A = \{1, 2, 3\}$, $B = \{x \in S : 2 \leq x \leq 7\}$, and $C = \{7, 8, 9, 10\}$.

   (a) Find $A \cup B$
   (b) Find $(A \cup C) - B$
   (c) Find $\bar{A} \cup (B - C)$
   (d) Do $A, B,$ and C form a partition of S?

Solution:
(a)

$$A \cup B = \{1, 2, 3, 4, 5, 6, 7\}$$

(b)

$$\begin{aligned} A \cup C &= \{1, 2, 3, 7, 8, 9, 10\} \\ B &= \{2, 3, \cdots, 7\} \\ \text{thus:} \quad (A \cup C) - B &= \{1, 8, 9, 10\} \end{aligned}$$

(c)

$$\begin{aligned} \bar{A} &= \{4, 5, \cdots, 10\} \\ B - C &= \{2, 3, 4, 5, 6\} \\ \text{thus:} \quad \bar{A} \cup (B - C) &= \{2, 3, \cdots, 10\} \end{aligned}$$

(d) No, since they are not disjoint. For example,

$$A \cap B = \{2, 3\} \neq \varnothing$$

3. For each of the following Venn diagrams, write the set denoted by the shaded area.

(a)

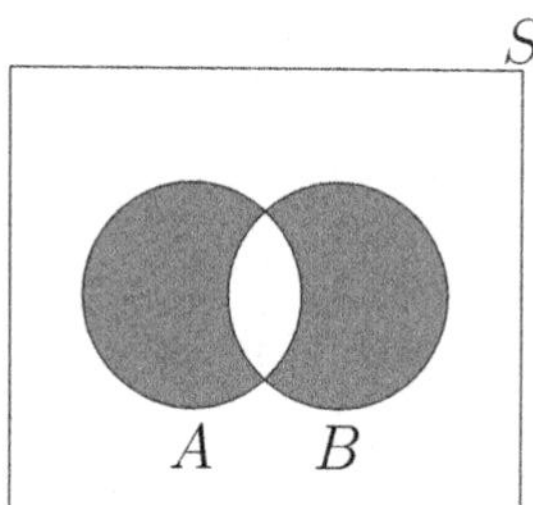

(b)

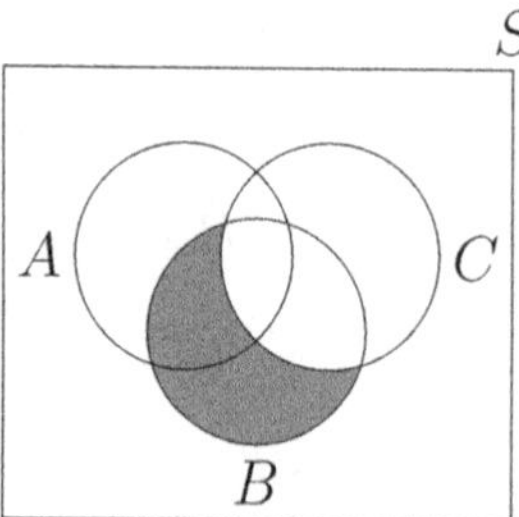

(c)

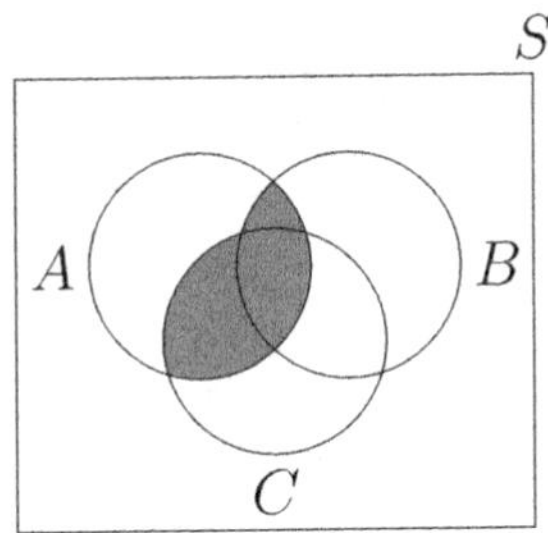

(d)

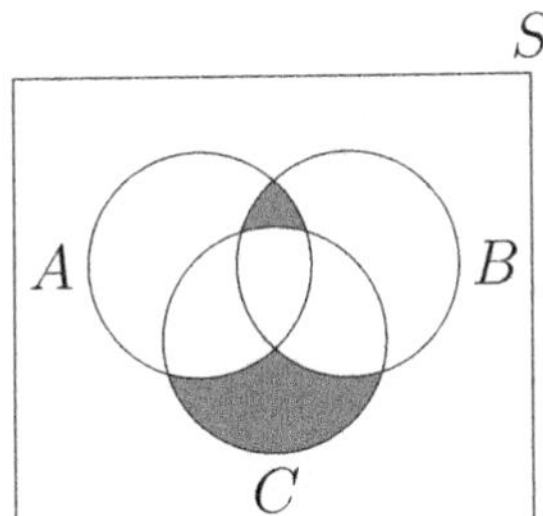

Solution: Note that there are generally several ways to represent each of the sets, so the answers to this question are not unique.

(a) $(A - B) \cup (B - A)$
(b) $B - C$
(c) $(A \cap B) \cup (A \cap C)$
(d) $(C - A - B) \cup ((A \cap B) - C)$

5. Let $A = \{1, 2, \cdots, 100\}$. For any $i \in \mathbb{N}$, define A_i as the set of numbers in A that are divisible by i. For example:

$A_2 = \{2, 4, 6, \cdots, 100\}$

$A_3 = \{3, 6, 9, \cdots, 99\}$
(a) Find $|A_2|$,$|A_3|$,$|A_4|$,$|A_5|$.
(b) Find $|A_2 \cup A_3 \cup A_5|$.

Solution:
(a) $|A_2| = 50$, $|A_3| = 33$, $|A_4| = 25$, $|A_5| = 20$.

Note that in general:
$|A_i| = \lfloor \frac{100}{i} \rfloor$, where $\lfloor x \rfloor$ is the largest integer less than or equal to x.

(b) By the inclusion-exclusion principle:

$$\begin{aligned} |A_2 \cup A_3 \cup A_5| = |A_2| + |A_3| + |A_5| \\ - |A_2 \cap A_3| - |A_2 \cap A_5| - |A_3 \cap A_5| \\ + |A_2 \cap A_3 \cap A_5|. \end{aligned}$$

We have:

$$\begin{aligned} |A_2| &= 50 \\ |A_3| &= 33 \\ |A_5| &= 20 \\ |A_2 \cap A_3| &= |A_6| = 16 \\ |A_2 \cap A_5| &= |A_{10}| = 10 \\ |A_3 \cap A_5| &= |A_{15}| = 6 \\ |A_2 \cap A_3 \cap A_5| &= |A_{30}| = 3 \\ |A_2 \cup A_3 \cup A_5| &= 50 + 33 + 20 \\ &- 16 - 10 - 6 \\ &+ 3 = 74 \end{aligned}$$

7. Determine whether each of the following sets is countable or uncountable.

(a) $A = \{1, 2, \cdots, 10^{10}\}$.

(b) $B = \{a + b\sqrt{2}| \quad a, b \in \mathbb{Q}\}$.
(c) $C = \{(X, Y) \in \mathbb{R}^2| \quad x^2 + y^2 \leq 1\}$.

Solution:
(a) A is countable because it is a finite set.
(b) B is countable because we can create a list with all the elements. Specifically, we have shown previously (refer to Figure 1.13 in the book) that if we can write any set B in the form of

$$B = \bigcup_i \bigcup_j \{q_{ij}\},$$

where indices i and j belong to some countable sets, that set in this form is countable.

For this case we can write

$$B = \bigcup_{i \in \mathbb{Q}} \bigcup_{j \in \mathbb{Q}} \{a_i + b_j\sqrt{2}\}.$$

So, we can replace q_{ij} by $a_i + b_j\sqrt{2}$.

(c) C is uncountable. To see this, note that for all $x \in [0, 1]$ then $(x, 0) \in C$.

9. Let $A_n = [0, \frac{1}{n}) = \{x \in \mathbb{R}| \quad 0 \leq x < \frac{1}{n}\}$ for $n = 1, 2, \cdots$. Define

$$A = \bigcap_{n=1}^{\infty} A_n = A_1 \cap A_2 \cap \cdots$$

Find A.

Solution:
By definition of the intersection

$$A = \{x | x \in A_n \quad \text{for } \underline{\text{all}} \quad n = 1, 2, \cdots\}$$

We claim $A = \{0\}$.
First note that $0 \in A_n$ for all $n = 1, 2, \cdots$. Thus $\{0\} \subset A$.

Next we show that A does not have any other elements. Since $A_n \subset [0,1)$ then $A \subset [0,1)$. Let $x \in (0,1)$. Choose $n > \frac{1}{x}$ then $\frac{1}{n} < x$. Thus $x \notin A_n$ and this results in $x \notin A$.

11. Show that the set $[0,1)$ is uncountable. That is, you can never provide a list in the form of $\{a_1, a_2, a_3, \cdots\}$ that contains all the elements in $[0,1)$.

Solution: Note that any $x \in [0,1)$ can be written in its binary expansion:

$$x = 0.b_1 b_2 b_3 \cdots$$

where $b_i \in \{0,1\}$. Now suppose that $\{a_1, a_2, a_3, \cdots\}$ is a list containing all $x \in [0,1)$. For example:

$$\begin{aligned} a_1 &= 0.\boxed{1}0101101001\cdots \\ a_2 &= 0.0\boxed{0}0110110111\cdots \\ a_3 &= 0.00\boxed{1}101001001\cdots \\ a_4 &= 0.100\boxed{1}001111001\cdots \end{aligned}$$

Now, we find a number $a \in [0,1)$ that does not belong to the list. Consider a such that the kth bit of a is the complement of the kth bit of a_k. For example, for the above list, a would be

$$a = 0.0100\cdots$$

We see that $a \notin \{a_1, a_2, \cdots\}$. This is a contradiction, so the above list cannot cover the entire $[0,1)$.

13. Two teams A and B play a soccer match, and we are interested in the winner. The sample space can be defined as:

$$S = \{a, b, d\}$$

where a shows the outcome that A wins, b shows the outcome that B wins, and d shows the outcome that they draw. Suppose that we know that (1) the probability that A wins is $P(a) = P(\{a\}) = 0.5$, and (2) the probability of a draw is $P(d) = P(\{d\}) = 0.25$.

(a) Find the probability that B wins.
(b) Find the probability that B wins or a draw occurs.

Solution:

$$\begin{aligned} P(a) + P(b) + P(d) &= 1 \\ P(a) &= 0.5 \\ P(d) &= 0.25 \end{aligned}$$

Therefore $P(b) = 0.25$.

(b)

$$\begin{aligned} P(\{b, d\}) &= P(b) + P(d) \\ &= 0.5 \end{aligned}$$

15. I roll a fair die twice and obtain two numbers. X_1 = result of the first roll, X_2 = result of the second roll.

(a) Find the probability that $X_2 = 4$.
(b) Find the probability that $X_1 + X_2 = 7$.
(c) Find the probability that $X_1 \neq 2$ and $X_2 \geq 4$.

Solution: The sample space has 36 elements:

$$\begin{aligned} S = \{&(1,1), (1,2), \cdots, (1,6), \\ &(2,1), (2,2), \cdots, (2,6), \\ &\vdots \\ &(6,1), (6,2), \cdots, (6,6)\} \end{aligned}$$

(a) The event $X_2 = 4$ can be represented by the set.

$$A = \{(1,4), (2,4), (3,4), (4,4), (5,4), (6,4)\}$$

Thus

$$P(A) = \frac{|A|}{|S|} = \frac{6}{36} = \frac{1}{6}$$

(b)

$$\begin{aligned} B &= \{(x_1, x_2)|x_1 + x_2 = 7\} \\ &= \{(1,6), (2,5), (3,4), (4,3), (5,2), (6,1)\} \end{aligned}$$

Therefore

$$P(B) = \frac{|6|}{36} = \frac{1}{6}$$

(c)

$$\begin{aligned} C &= \{(X_1, X_2)|X_1 \neq 2, X_2 \geq 4\} \\ &= \{(1,4), (1,5), (1,6), \\ &(3,4), (3,5), (3,6), \\ &(4,4), (4,5), (4,6), \\ &(5,4), (5,5), (5,6), \\ &(6,4), (6,5), (6,6)\} \end{aligned}$$

Therefore

$$|C| = 15$$

Which results in:

$$P(C) = \frac{15}{36} = \frac{5}{12}.$$

17. Four teams A, B, C, and D compete in a tournament. Teams A and B have the same chance of winning the tournament. Team C is twice as likely to win the tournament as team D. The probability that either team A or team C wins the tournament is 0.6. Find the probabilities of each team winning the tournament.

Solution: We have

$$\begin{cases} P(A) = P(B) \\ P(C) = 2P(D) \\ P(A \cup C) = 0.6 \qquad\qquad \text{thus} \quad P(A) + P(C) = 0.6 \\ P(A) + P(B) + P(C) + P(D) = 1 \end{cases}$$

which results in

$$\begin{cases} P(A) = P(B) = P(D) = 0.2 \\ P(C) = 0.4 \end{cases}$$

19. You choose a point (A, B) uniformly at random in the unit square $\{(x, y) : 0 \leq x, y \leq 1\}$.

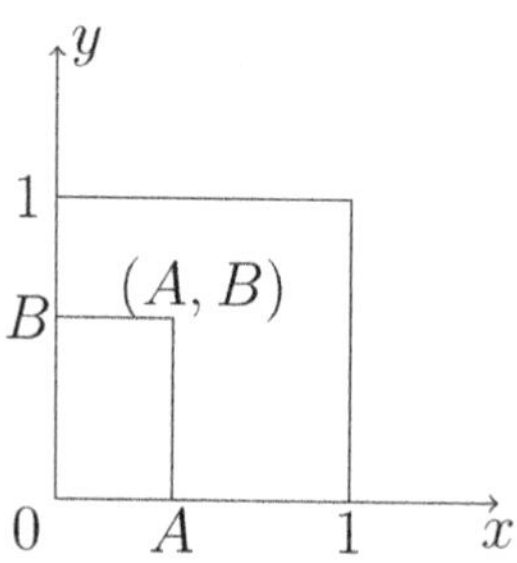

What is the probability that the equation

$$AX^2 + X + B = 0$$

has real solutions?

Solution: The equation has real roots if and only if:

$$1 - 4AB \geq 0 \quad \text{i.e.} \quad AB \leq \frac{1}{4}.$$

This area is shown here:

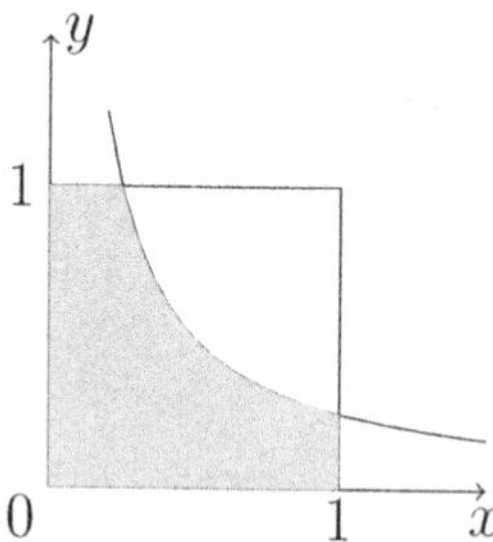

Since (A, B) is uniformly chosen in the square, we can say that the probability of having real roots is

$$\begin{aligned} P(R) &= \frac{\text{area of the shaded region}}{\text{area of the square}} \\ &= \frac{\text{area of the shaded region}}{1} \end{aligned}$$

To find the area of the shaded region we can set up the following integral:

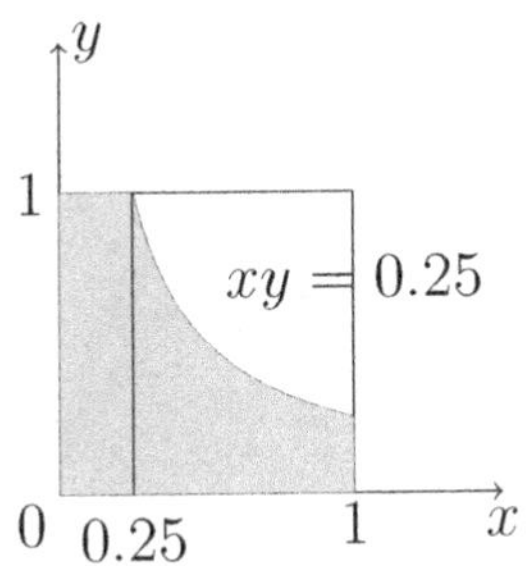

$$
\begin{aligned}
Area &= \frac{1}{4} + \int_{\frac{1}{4}}^{1} \frac{1}{4x} dx \\
&= \frac{1}{4} + \frac{1}{4} \left[\ln(x)\right]_{\frac{1}{4}}^{1} \\
&= \frac{1}{4} + \frac{1}{4} \ln 4
\end{aligned}
$$

21. (continuity of probability) For any sequence of events $A_1, A_2, A_3, \cdots$. Prove

$$
P\left(\bigcup_{i=1}^{\infty} A_i\right) = \lim_{n\to\infty} P\left(\bigcup_{i=1}^{n} A_i\right)
$$

$$
P\left(\bigcap_{i=1}^{\infty} A_i\right) = \lim_{n\to\infty} P\left(\bigcap_{i=1}^{n} A_i\right)
$$

Solution: Define the new sequence $B_1, B_2, \cdots$ as

$$
\begin{aligned}
B_1 &= A_1 \\
B_2 &= A_2 - A_1 \\
B_3 &= A_3 - (A_1 \cup A_2) \\
&\vdots \\
B_i &= A_i - \left(\bigcup_{j=1}^{i-1} A_j\right)
\end{aligned}
$$

Then we have:

(a) B_i's are disjoint.
(b) $\bigcup_{i=1}^{n} B_i = \bigcup_{i=1}^{n} A_i$.
(c) $\bigcup_{i=1}^{\infty} B_i = \bigcup_{i=1}^{\infty} A_i$.

Then we can write:

$$
\begin{aligned}
P\left(\bigcup_{i=1}^{\infty} A_i\right) &= P\left(\bigcup_{i=1}^{\infty} B_i\right) \\
&= \sum_{i=1}^{\infty} P(B_i) \quad (Bi\text{'s are disjoint}) \\
&= \lim_{n\to\infty}\left(\sum_{i=1}^{n} P(B_i)\right) \quad (\text{definition of infinite sum}) \\
&= \lim_{n\to\infty}\left[P\left(\bigcup_{i=1}^{n} B_i\right)\right] \quad (Bi\text{'s are disjoint}) \\
&= \lim_{n\to\infty}\left[P\left(\bigcup_{i=1}^{n} A_i\right)\right]
\end{aligned}
$$

To prove the second part, apply the result of the first part to $A_1^c, A_2^c, \cdots$.
Note: You can also solve this problem using what you have already shown in Problem 20.

23. Let A, B, and C be three events with probabilities given:

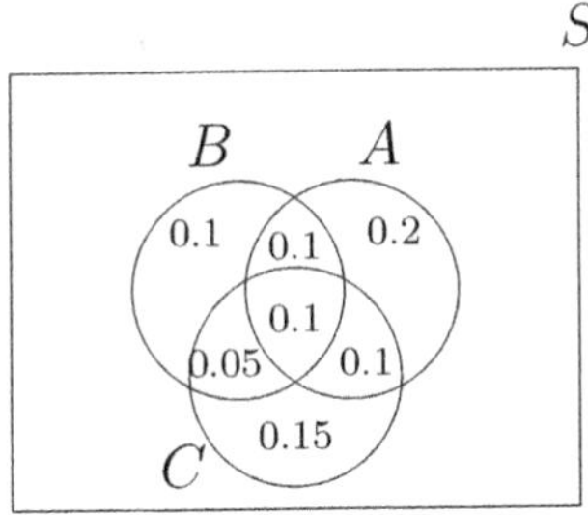

(a) Find $P(A|B)$
(b) Find $P(C|B)$
(c) Find $P(B|A \cup C)$
(d) Find $P(B|A, C) = P(B|A \cap C)$

Solution:

(a)

$$\begin{aligned} P(A|B) &= \frac{P(A \cap B)}{P(B)} \\ &= \frac{0.2}{0.35} \\ &= \frac{4}{7} \end{aligned}$$

(b)

$$\begin{aligned} P(C|B) &= \frac{P(C \cap B)}{P(B)} \\ &= \frac{0.15}{0.35} \\ &= \frac{3}{7} \end{aligned}$$

(c)

$$
\begin{aligned}
P(B|A \cup C) &= \frac{P(B \cap (A \cup C))}{P(A \cup C)} \\
&= \frac{0.1 + 0.1 + 0.05}{0.2 + 0.1 + 0.1 + 0.1 + 0.5 + 0.05} \\
&= \frac{0.25}{0.7} \\
&= \frac{5}{14}
\end{aligned}
$$

(d)

$$
\begin{aligned}
P(B|A, C) &= \frac{P(B \cap A \cap C)}{P(A \cap C)} \\
&= \frac{0.1}{0.2} \\
&= \frac{1}{2}
\end{aligned}
$$

25. A professor thinks students who live on campus are more likely to get As in the probability course. To check this theory, the professor combines the data from the past few years:

    1. 600 students have taken the course.
    2. 120 students have got As.
    3. 200 students lived on campus.
    4. 80 students lived off campus and got As.

    Does this data suggest that "getting an A" and "living on campus" are dependent or independent?

Solution: From the data, you can see that 80 students out of the 400 off-campus students got an A (20%). Also, 40 students out of the 200 on-campus students got an A (again 20%). Thus, the data suggests that "getting an A" and "living on campus" are independent. You can also see this using the definitions of independence in the following way:

Let C be the event that a random student lives on campus and A be the event that he or she gets an A in the course. We have:

$$\begin{aligned}
P(A) &\approx \frac{120}{600} = \frac{1}{5} \\
P(C) &\approx \frac{200}{600} = \frac{1}{3} \\
P(A \cap C^c) &\approx \frac{80}{600} = \frac{2}{15} \\
P(A \cap C) &= P(A) - P(A \cap C^c) \\
&= \frac{1}{5} - \frac{2}{15} \\
&= \frac{1}{15}
\end{aligned}$$

Therefore,

$$\begin{aligned}
\frac{1}{15} &= P(A \cap C) \\
&= P(A).P(C)
\end{aligned}$$

The data suggests that A and C are independent.

27. Consider a communication system. At any given time, the communication channel is in good condition with probability 0.8 and is in bad condition with probability 0.2. An error occurs in a transmission with probability 0.1 if the channel is in good condition and with probability 0.3 if the channel is in bad condition. Let G be the event that the channel is in good condition and E be the event that there is an error in transmission.

   (a) Complete the following tree diagram:

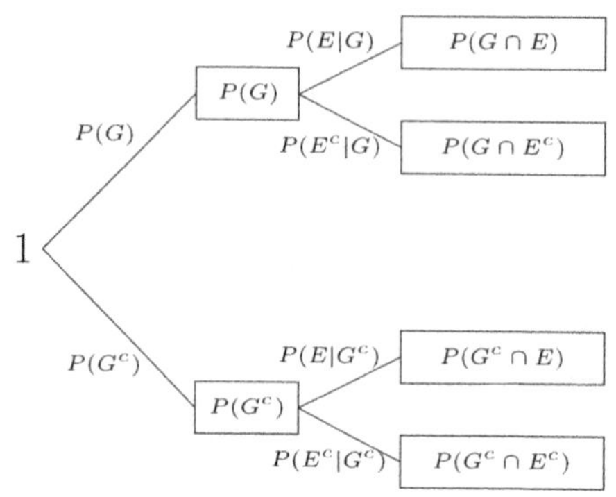

(b) Using the tree find $P(E)$.
(c) Using the tree find $P(G|E^c)$.

Solution:

(a)

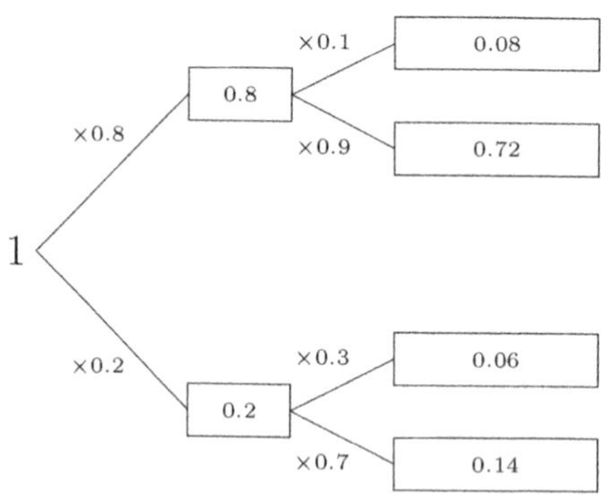

(b)

$$\begin{aligned} P(E) &= P(G \cap E) + P(G^c \cap E) \\ &= 0.08 + 0.06 \\ &= 0.14 \end{aligned}$$

(c)

$$\begin{aligned} P(G|E^c) &= \frac{P(G \cap E^c)}{P(E^c)} \\ &= \frac{0.72}{1 - 0.14} \\ &= \frac{0.72}{0.86} \\ &\approx 0.84 \end{aligned}$$

29. Reliability:
Real-life systems often are comprised of several components. For example, a system may consist of two components that are connected in parallel as shown in Figure 1.1. When the system's components are connected in parallel, the system works if at least one of the components is functional. The components might also be connected in series as shown in Figure 1.1. When the system's components are connected in series, the system works if all of the components are functional.

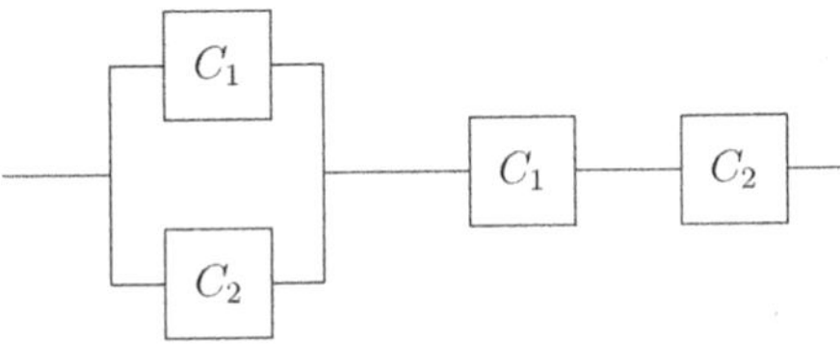

Figure 1.1: In the left figure, Components C_1 and C_2 are connected in parallel. The system is functional if at least one of the C_1 and C_2 is functional. In the right figure, Components C_1 and C_2 are connected in series. The system is functional only if both C_1 and C_2 are functional.

For each of the following systems, find the probability that the system is functional. Assume that component k is functional with probability P_k independent of other components.

(a)

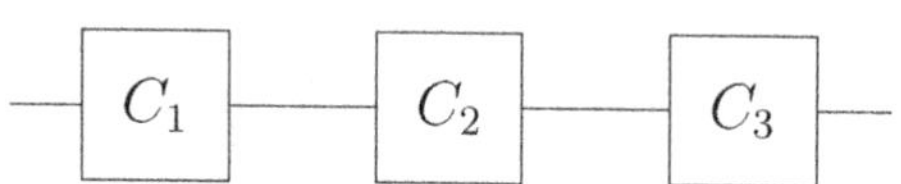

(b)

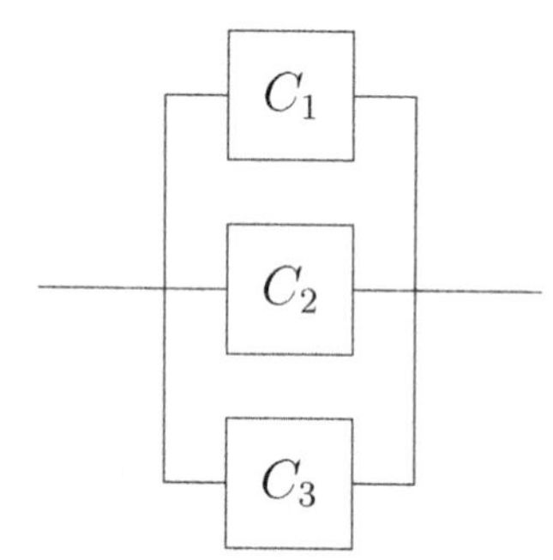

(c)

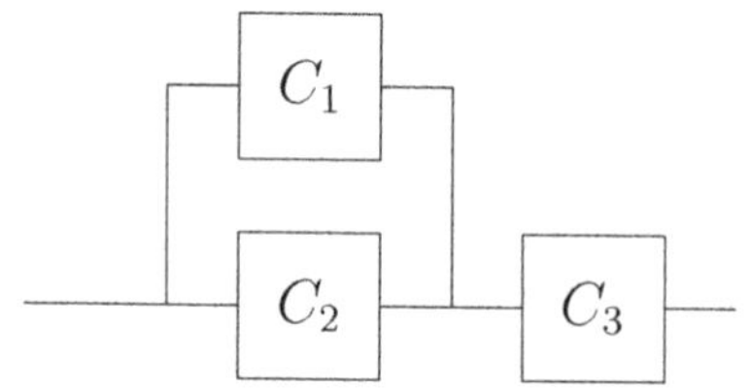

(d)

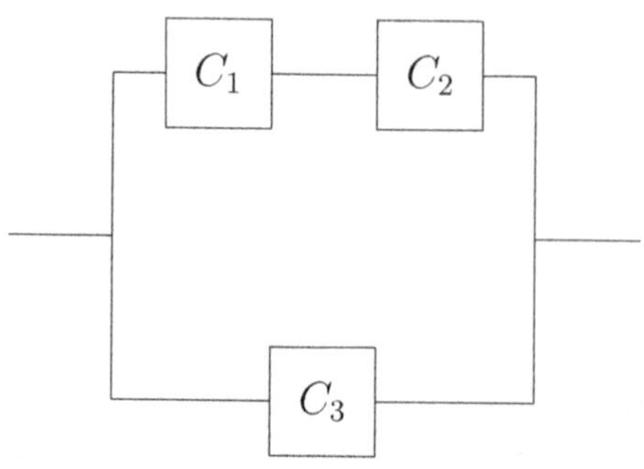

(e)

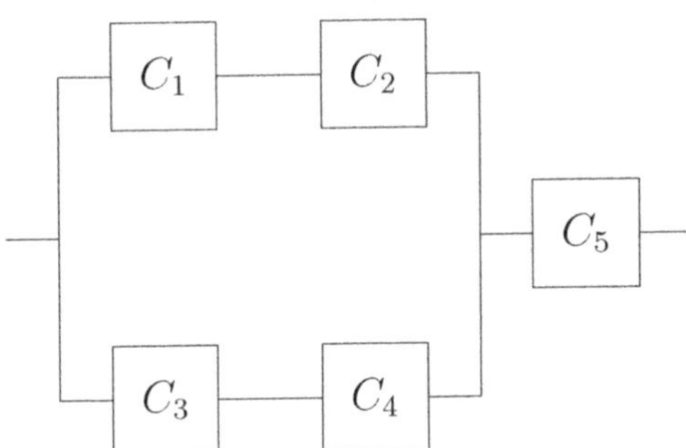

Solution:

Let A_k be the event that the kth component is functional and let A be the event that the whole system is functional.

(a)

$$\begin{aligned} P(A) &= P(A_1 \cap A_2 \cap A_3) \\ &= P(A_1) \cdot P(A_2) \cdot P(A_3) \quad \text{(since } A_i\text{s are independent)} \\ &= P_1 P_2 P_3 \end{aligned}$$

(b)

$$\begin{aligned} P(A) &= P(A_1 \cup A_2 \cup A_3) \\ &= 1 - P(A_1^c \cap A_2^c \cap A_3^c) \quad \text{(Demorgan's law)} \\ &= 1 - P(A_1^c)P(A_2^c)P(A_3^c) \quad \text{(since } A_i\text{s are independent)} \\ &= 1 - (1 - P_1)(1 - P_2)(1 - P_3). \end{aligned}$$

(c)

$$\begin{aligned} P(A) &= P((A_1 \cup A_2) \cap A_3) \\ &= P(A_1 \cup A_2) \cdot P(A_3) \quad \text{(since } A_i\text{s are independent)} \\ &= [1 - P(A_1^c \cap A_2^c)] \cdot P(A_3) \\ &= [1 - (1 - P_1)(1 - P_2)]P_3 \end{aligned}$$

(d)

$$
\begin{aligned}
P(A) &= P\left[(A_1 \cap A_2) \cup A_3\right] \\
&= 1 - P\left((A_1 \cap A_2)^c\right) \cdot P(A_3^c) \quad \text{(since } A_i\text{s are independent)} \\
&= 1 - (1 - P(A_1) \cdot P(A_2))\,(1 - P(A_3)) \\
&= 1 - (1 - P_1P_2)(1 - P_3)
\end{aligned}
$$

(e)

$$
\begin{aligned}
P(A) &= P\left[((A_1 \cap A_2) \cup (A_3 \cap A_4)) \cap A_5\right] \\
&= P\left((A_1 \cap A_2) \cup (A_3 \cap A_4)\right) \cdot P(A_5) \quad \text{(since } A_i\text{s are independent)} \\
&= \left[1 - (1 - P(A_1 \cap A_2)) \cdot (1 - P(A_3 \cap A_4))\right] P_5 \quad \text{(parallel links)} \\
&= \left[1 - (1 - P_1P_2)(1 - P_3P_4)\right] P_5
\end{aligned}
$$

31. One way to design a spam filter is to look at the words in an email. In particular, some words are more frequent in spam emails. Suppose that we have the following information:

    1. 50% of emails are spam.
    2. 1% of spam emails contain the word "refinance."
    3. 0.001% of non-spam emails contain the word "refinance."

    Suppose that an email is checked and found to contain the word refinance. What is the probability that the email is spam?

    Solution:

    Let S be the event that an email is spam and let R be the event that the email contains the word "refinance." Then,

$$
\begin{aligned}
P(S) &= \frac{1}{2} \\
P(R|S) &= \frac{1}{100} \\
P(R|S^c) &= \frac{1}{100000}
\end{aligned}
$$

Then,

$$\begin{aligned}
P(S|R) &= \frac{P(R|S)P(S)}{P(R)} \\
&= \frac{P(R|S)P(S)}{P(R|S)P(S) + P(R|S^c)P(S^c)} \\
&= \frac{\frac{1}{100} \times \frac{1}{2}}{\frac{1}{100} \times \frac{1}{2} + \frac{1}{100000} \times \frac{1}{2}} \\
&\approx 0.999
\end{aligned}$$

33. (The Monte Hall Problem[1]) You are in a game show, and the host gives you the choice of three doors. Behind one door is a car and behind the others are goats. Say you pick door 1. The host, who knows what is behind the doors, opens a different door and reveals a goat (the host can always open such a door because there is only one door with a car behind it). The host then asks you: "Do you want to switch?" The question is, is it to your advantage to switch your choice?

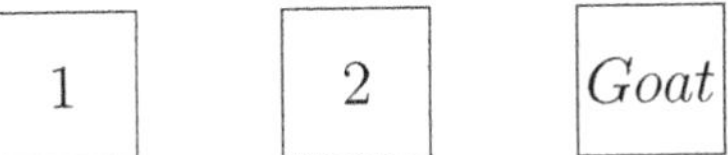

Solution: Yes, if you switch, your chance of winning the car is $\frac{2}{3}$. Let W be the event that you win the car if you switch. Let C_i be the event that the car is behind door i, for $i = 1, 2, 3$. Then $P(C_i) = \frac{1}{3} \quad i = 1, 2, 3$. Note that if the car is behind either door 2 or 3 you will win by switching, so $P(W|C_2) = P(W|C_3) = 1$. On the other hand, if the car is behind door 1 (the one you originally chose), you will lose by switching, so $P(W|C_1) = 0$.

[1] http://en.wikipedia.org/wiki/Monty_Hall_problem

Then,

$$\begin{aligned}
P(W) &= \sum_{i=1}^{3} P(W|C_i)P(C_i) \\
&= P(W|C_1)P(C_1) + P(W|C_2)P(C_2) + P(W|C_3)P(C_3) \\
&= 0 \cdot \frac{1}{3} + 1 \cdot \frac{1}{3} + 1 \cdot \frac{1}{3} \\
&= \frac{2}{3}.
\end{aligned}$$

35. You and I play the following game: I toss a coin repeatedly. The coin is unfair and $P(H) = p$. The game ends the first time that two consecutive heads (HH) or two consecutive tails (TT) are observed. I win if (HH) is observed and you win if (TT) is observed. Given that I won the game, find the probability that the first coin toss resulted in head.

Solution:

Let A be the event that I win.

$$P(A) = P(A|H)P(H) + P(A|T)P(T)$$

$P(A|H)$: the probability that I win given that the first coin toss is a head.

$$\begin{aligned}
A|H &: HH, HTHH, HTHTHH, \cdots \\
P(A|H) &= p + pqp + (pq)^2 p + \cdots \\
&= p[1 + pq + \cdots] \\
&= \frac{p}{1 - pq}.
\end{aligned}$$

$$\begin{aligned}
A|T &: THH, THTHH, THTHTHH, \cdots \\
P(A|T) &= p^2 + p(1-p)p^2 + \cdots \\
&= p^2[1 + pq + (pq)^2 + \cdots] \\
&= \frac{p^2}{1-pq} \\
P(A) &= P(A|H)P(H) + P(A|T)P(T) \\
&= \frac{p^2}{1-pq} + \frac{p^2 q}{1-pq} \\
&= \frac{p^2(1+q)}{1-pq}.
\end{aligned}$$

$$\begin{aligned}
P(H|A) &= \frac{P(A|H)P(H)}{P(A)} \\
&= \frac{\frac{p^2}{1-pq}}{\frac{p^2}{1-pq}(1+q)} \\
&= \frac{1}{1+q} \\
&= \frac{1}{2-p}
\end{aligned}$$

37. A family has n children, $n \geq 2$. What is the probability that all children are girls, given that at least one of them is a girl?

Solution:

The sample space has 2^n elements,

$$S = \{(G, G, \cdots, G), (G, \cdots, B), \cdots, (B, B, \cdots, B)\}.$$

Let A be the event that all the children are girls, then

$$A = \{(G, G, \cdots, G)\}.$$

Thus

$$P(A) = \frac{1}{2^n}.$$

Let B be the event that at least one child is a girl, then:

$$\begin{aligned} B &= S - \{(B, \cdots, B)\} \\ |B| &= 2^n - 1 \\ P(B) &= \frac{2^n - 1}{2^n}. \end{aligned}$$

Then

$$\begin{aligned} A \cap B &= A \\ P(A|B) &= \frac{P(A \cap B)}{P(B)} \\ &= \frac{P(A)}{P(B)} \\ &= \frac{\frac{1}{2^n}}{\frac{2^n-1}{2^n}} \\ &= \frac{1}{2^n - 1} \end{aligned}$$

Note: If we let $n = 2$, we obtain $P(A|B) = \frac{1}{3}$ which is the same as Example 17 in the text.

39. A family has n children. We pick one of them at random and find out that she is a girl. What is the probability that all their children are girls?

Solution:

Let Gr be the event that a randomly chosen child is a girl. Let A be the event that all the children are girls. Then,

$$\begin{aligned} P(Gr|A) &= 1 \\ P(A) &= \frac{1}{2^n} \\ P(Gr) &= \frac{1}{2} \end{aligned}$$

Thus,

$$
\begin{aligned}
P(A|Gr) &= \frac{P(Gr|A)P(A)}{P(Gr)} \\
&= \frac{1 \cdot \frac{1}{2^n}}{\frac{1}{2}} \\
&= \frac{1}{2^{n-1}}
\end{aligned}
$$

Chapter 2

Combinatorics: Counting Methods

1. A coffee shop has 4 different types of coffee. You can order your coffee in a small, medium, or large cup. You can also choose whether you want to add cream, sugar, or milk (any combination is possible. For example, you can choose to add all three). In how many ways can you order your coffee?

   Solution:

   We can use the multiplication principle to solve this problem. There are 4 choices for the coffee type, 3 choices for the cup size, 2 choices for cream (adding cream or no cream), 2 choices for sugar, and 2 choices for milk. Thus, the total number of ways we can order our coffee is equal to:

$$4 \times 3 \times 2 \times 2 \times 2 = 96$$

3. There are 20 black cell phones and 30 white cell phones in a store. An employee takes 10 phones at random. Find the probability that

(a) there will be exactly 4 black cell phones among the chosen phones.

(b) there will be less than 3 black cell phones among the chosen phones.

Solution:

(a) Let A be the event that there are exactly 4 black cell phones among the 10 chosen cell phones. Then:

$$P(A) = \frac{|A|}{|S|}$$

$$|S| = \binom{50}{10}$$

$$|A| = \binom{20}{4}\binom{30}{6}$$

Thus:

$$P(A) = \frac{\binom{20}{4}\binom{30}{6}}{\binom{50}{10}}.$$

(b) Let B be the event that there are less than 3 black cell phones among the chosen phones. Then:

$$\begin{aligned} P(B) &= P(\text{“0 black phones” or “1 black phones” or “2 black phones”}) \\ &= \frac{\binom{20}{0}\binom{30}{10} + \binom{20}{1}\binom{30}{9} + \binom{20}{2}\binom{30}{8}}{\binom{50}{10}} \end{aligned}$$

5. Five cards are dealt from a shuffled deck. What is the probability that the hand contains exactly two aces, given that we know it contains at least one ace?

Solution:

Let A be the event that the hand contains exactly two aces and B the event that it contains at least one ace.

We can use the formula for the conditional probability:

$$\begin{aligned} P(A|B) &= \frac{P(A \cap B)}{P(B)} \\ &= \frac{P(A)}{P(B)} = \frac{P(A)}{1 - P(B^c)} \end{aligned}$$

$$P(A) = \frac{\binom{4}{2}\binom{48}{3}}{\binom{52}{5}}$$

$$P(B^c) = \frac{\binom{48}{5}}{\binom{52}{5}}$$

By substituting $P(A)$ and $P(B^c)$ to the equation of $P(A|B)$, we have:

$$\begin{aligned} P(A|B) &= \frac{\frac{\binom{4}{2}\binom{48}{3}}{\binom{52}{5}}}{1 - \frac{\binom{48}{5}}{\binom{52}{5}}} \\ &= \frac{\binom{4}{2}\binom{48}{3}}{\binom{52}{5} - \binom{48}{5}} \end{aligned}$$

7. There are 50 students in a class and the professor chooses 15 students at random. What is the probability that neither you nor your friend Joe are among the chosen students?

Solution:

There are 50 students. A is the event that you or Joe are among the 15 chosen students. We can consider the following simplification:

$$50 \text{ students} = \text{you} + \text{your friend Joe} + 48 \text{ others}$$

We can solve the problem by calculating $P(A^c)$. A^c is the event that neither you or your friend Joe is selected. Thus:

$$\begin{aligned} P(A) &= 1 - P(A^c) \\ &= 1 - \frac{\binom{48}{15}}{\binom{50}{15}} \end{aligned}$$

9. You have a biased coin for which $P(H) = p$. You toss the coin 20 times. What is the probability that:

   (a) You observe 8 heads and 12 tails?

   (b) You observe more than 8 heads and more than 8 tails?

Solution:

(a) Let A be the event that you observe 8 heads and 12 tails. For this problem we can use the binomial formula:

$$P(8 \text{ heads}) = \binom{20}{8} p^8 (1-p)^{12}.$$

(b) Let X be the number of heads and Y be the number of tails. Because you toss the coin 20 times, $X + Y = 20$.

Let B be the event that you observe more than 8 heads and more than 8 tails. Then:

$$\begin{aligned} P(B) &= P(X > 8 \text{ and } Y > 8) \\ &= P((X > 8) \text{ and } (20 - X > 8)) \\ &= P(8 < X < 12) \\ &= \sum_{k=9}^{11} \binom{20}{k} p^k (1-p)^{20-k} \end{aligned}$$

11. In problem 10, assume that all the appropriate paths are equally likely. What is the probability that the sensor located at point $(10, 5)$ receives the message (that is, what is the probability that a randomly chosen path from $(0, 0)$ to $(20, 10)$ goes through the point $(10, 5)$)?

Solution:

We need to count the number of paths going from $(0, 0)$ to $(20, 10)$ that go through the point $(10, 5)$. The number of such paths is equal to the number of paths from $(0, 0)$ to $(10, 5)$ multiplied by the number of paths from $(10, 5)$ to $(20, 10)$ which is equal to

$$\binom{15}{5} \times \binom{15}{5} = \binom{15}{5}^2.$$

Let A be the event that the sensor located at point $(10, 5)$ receives the message. Thus:

$$P(A) = \frac{\binom{15}{5}^2}{\binom{30}{10}}$$

13. There are two coins in a bag. For coin 1, $P(H) = \frac{1}{2}$ and for coin 2, $P(H) = \frac{1}{3}$. Your friend chooses one of the coins at random and tosses it 5 times.

    (a) What is the probability of observing at least 3 heads?

    (b) You ask your friend, "did you observe at least three heads?" Your friend replies, "yes." What is the probability that coin 2 was chosen?

Solution:

(a) Let A be the event that your friend observes at least 3 heads. If we know the value of $P(H)$, then $P(A)$ is given by

$$P(A) = \sum_{k=3}^{5} \binom{5}{k} P(H)^k (1 - P(H))^{5-k}.$$

Thus,

$$P(A|\text{coin1}) = \sum_{k=3}^{5} \binom{5}{k} (\frac{1}{2})^5,$$

and

$$P(A|\text{coin2}) = \sum_{k=3}^{5} \binom{5}{k} (\frac{1}{3})^k (\frac{2}{3})^{(5-k)}.$$

Using the law of total probability,

$$\begin{aligned} P(A) &= P(A|\text{coin1}).P(\text{coin1}) + P(A|\text{coin2}).P(\text{coin2}) \\ &= \left(\sum_{k=3}^{5} \binom{5}{k} (\frac{1}{2})^5\right) \cdot \frac{1}{2} + \left(\sum_{k=3}^{5} \binom{5}{k} (\frac{1}{3})^k (\frac{2}{3})^{(5-k)}\right) \cdot \frac{1}{2} \end{aligned}$$

(b)

$$
\begin{aligned}
P(\text{coin2}|A) &= \frac{P(A|\text{coin2}).P(\text{coin2})}{P(A)} \\
&= \frac{\left(\sum_{k=3}^{5}\binom{5}{k}\left(\frac{1}{3}\right)^k\left(\frac{2}{3}\right)^{(5-k)}\right)}{\left(\sum_{k=3}^{5}\binom{5}{k}\left(\frac{1}{2}\right)^5\right)+\left(\sum_{k=3}^{5}\binom{5}{k}\left(\frac{1}{3}\right)^k\left(\frac{2}{3}\right)^{(5-k)}\right)}
\end{aligned}
$$

15. You roll a die 5 times. What is the probability that at least one value is observed more than once?

Solution:

Let A be the event that at least one value is observed more than once. Then, A^c is the event in which no repetition is observed.

$$
\begin{aligned}
P(A^c) &= \frac{|A^c|}{|S|} \\
&= \frac{6 \times 5 \times 4 \times 3 \times 2}{6^5} \\
&= \frac{5}{54}
\end{aligned}
$$

So, we can conclude:

$$
P(A) = 1 - \frac{5}{54} = \frac{49}{54}
$$

17. I have have two bags. Bag 1 contains 10 blue marbles, while bag 2 contains 15 blue marbles. I pick one of the bags at random, and throw 6 red marbles in it. Then I shake the bag and choose 5 marbles (without replacement) at random from the bag. If there are exactly 2 red marbles among the 5 chosen marbles, what is the probability that I have chosen bag 1?

Solution:

We have the following information:

Bag 1: 10 blue marbles.

Bag 2: 15 blue marbles.

Let A be the event that exactly 2 red marbles among the 5 chosen marbles exist. Let B_1 be the event that Bag 1 has been chosen. Let B_2 be the event that Bag 2 has been chosen.

We want to calculate $P(B_1|A)$. We use Bayes' rule:

$$\begin{aligned} P(B_1|A) &= \frac{P(A|B_1)P(B_1)}{P(A)} \\ &= \frac{P(A|B_1)P(B_1)}{P(A|B_1)P(B_1) + P(A|B_2)P(B_2)} \end{aligned}$$

First, note that $P(B_1) = P(B_2) = \frac{1}{2}$. If Bag 1 is chosen, there will be 10 blue and 6 red marbles in the bag, so the probability of choosing two red marbles will be

$$P(A|B_1) = \frac{\binom{6}{2}\binom{10}{3}}{\binom{16}{5}}.$$

Similarly,

$$P(A|B_2) = \frac{\binom{6}{2}\binom{15}{3}}{\binom{21}{5}}$$

Thus:

$$P(B_1|A) = \frac{\frac{\binom{6}{2}\binom{10}{3}}{\binom{16}{5}}}{\frac{\binom{6}{2}\binom{10}{3}}{\binom{16}{5}} + \frac{\binom{6}{2}\binom{15}{3}}{\binom{21}{5}}}$$
$$= \frac{\binom{21}{5}\binom{10}{3}}{\binom{21}{5}\binom{10}{3} + \binom{15}{3}\binom{16}{5}}$$

19. How many distinct solutions does the following equation have such that all $x_i \in \mathbb{N}$?

$$x_1 + x_2 + x_3 + x_4 + x_5 = 100$$

Solution:

Define $y_i = x_i - 1$, then $y_i \in \{0, 1, 2, \cdots\}$. We can rewrite the equations as:

$$(y_1 + 1) + (y_2 + 1) + (y_3 + 1) + (y_4 + 1) + (y_5 + 1) = 100$$
$$\text{such that } y_i \in \{0, 1, 2, \cdots\}$$

So, we conclude:

$$y_1 + y_2 + y_3 + y_4 + y_5 = 95 \text{ such that } y_i \in \{0, 1, 2, \cdots\}$$

Thus, using Theorem 2.1 in the textbook, the number of the solutions is:

$$\binom{95 + 5 - 1}{5 - 1} = \binom{99}{4}.$$

21. For this problem, suppose that x_i's must be non-negative integers, i.e., $x_i \in \{0, 1, 2, \cdots\}$ for $i = 1, 2, 3$. How many distinct solutions does the following equation have such that at least one of the x_i's is larger than 40?

$$x_1 + x_2 + x_3 = 100$$

Solution:

Let A_i be the set of solutions to $x_1 + x_2 + x_3 = 100, x_i \in \{0, 1, 2, \cdots\}$ for $i = 1, 2, 3$ such that $x_i > 40$. Then by the inclusion-exclusion principle:

$$\begin{aligned} |A_1 \cup A_2 \cup A_3| &= |A_1| + |A_2| + |A_3| \\ &\quad - |A_1 \cap A_2| - |A_1 \cap A_3| - |A_2 \cap A_3| \\ &\quad + |A_1 \cap A_2 \cap A_3| \\ &= 3|A_1| - 3|A_1 \cap A_2| + |A_1 \cap A_2 \cap A_3| \end{aligned}$$

Note that we used the fact that by symmetry, we have

$$\begin{aligned} |A_1| + |A_2| + |A_3| &= 3|A_1| \\ |A_1 \cap A_2| + |A_1 \cap A_3| + |A_2 \cap A_3| &= 3|A_1 \cap A_2|. \end{aligned}$$

To find $|A_1|$:

$$y_1 = x_1 - 41$$

Thus, $y_1 \in \{0, 1, 2, \cdots\}$. We want to find the number of the solutions to the equation: $y_1 + x_2 + x_3 = 59, y_1, x_2, x_3 \in \{0, 1, 2, \cdots\}$.

Thus:

$$|A_1| = \binom{59 + 3 - 1}{3 - 1} = \binom{61}{2}.$$

To find $|A_1 \cap A_2|$:

define:

$$y_1 = x_1 - 41$$
$$y_2 = x_2 - 41$$

So, we have:

$$y_1 + y_2 + x_3 = 18, \text{ such that } y_1, y_2, x_3 \in \{0, 1, 2, \cdots\}$$

We get:

$$|A_1 \cap A_2| = \binom{18+3-1}{3-1} = \binom{20}{2}.$$

To find $|A_1 \cap A_2 \cap A_3|$:

define:

$$y_i = x_i - 41 \text{ for } i = 1, 2, 3$$

This event cannot happen for $x_i > 40$ $i = 1, 2, 3$, we have $x_1 + x_2 + x_3 > 120$. So this equation $x_1 + x_2 + x_3 = 100$ does not have any solution for $x_i > 40$ $i = 1, 2, 3$.

So, we have:

$$|A_1 \cap A_2 \cap A_3| = 0$$

Thus:

$$|A_1 \cup A_2 \cup A_3| = 3\binom{61}{2} - 3\binom{20}{2} = 4920.$$

There is also another way to solve this problem. We find the number of solutions in which none of x_i's are greater than 40. In other words, all x_i's $\in 0, 1, 2, ..., 40$ for $i = 1, 2, 3$

We define $y_i = 40 - x_i$ for $i = 1, 2, 3$.

We want $y_i \geq 0$, and $x_i \in 0, 1, 2, ..., 40$.

$x_1 + x_2 + x_3 = 100, x_i \in \{0, 1, 2, \cdots\}$ for $i = 1, 2, 3$ such that $x_i \leq 40$.

$40 - y_1 + 40 - y_2 + 40 - y_3 = 100, y_i \in \{0, 1, 2, \cdots, 40\}$

$$y_1 + y_2 + y_3 = 20, \text{ such that } y_1, y_2, y_3 \in \{0, 1, 2, \cdots\}$$

The number of solutions is:

$$\binom{20+3-1}{3-1} = \binom{22}{2}.$$

So, the number of solutions in which at least one of the x_i's is greater than 40 is equal to the total number of solutions minus $\binom{22}{2}$. Using Theorem 2.1, the total number of solutions is

$$\binom{102}{2}.$$

Thus, the number of solutions in which at least one of the x_i's is greater than 40 is equal to

$$\binom{102}{2} - \binom{22}{2} = 4920.$$

Chapter 3

Discrete Random Variables

1. Let X be a discrete random variable with the following PMF

$$P_X(x) = \begin{cases} \frac{1}{2} & \text{for } x = 0 \\ \frac{1}{3} & \text{for } x = 1 \\ \frac{1}{6} & \text{for } x = 2 \\ 0 & \text{otherwise} \end{cases}$$

   (a) Find R_X, the range of the random variable X.
   (b) Find $P(X \geq 1.5)$.
   (c) Find $P(0 < X < 2)$.
   (d) Find $P(X = 0|X < 2)$

Solution:

(a) The range of X can be found from the PMF. The range of X consists of possible values for X. Here we have

$$R_X = \{0, 1, 2\}.$$

(b) The event $X \geq 1.5$ can happen only if X is 2. Thus,

$$\begin{aligned} P(X \geq 1.5) &= P(X = 2) \\ &= P_X(2) = \frac{1}{6}. \end{aligned}$$

(c) Similarly, we have

$$\begin{aligned} P(0 < X < 2) &= P(X = 1) \\ &= P_X(1) = \frac{1}{3}. \end{aligned}$$

(d) This is a conditional probability problem, so we can use our famous formula $P(A|B) = \frac{P(A \cap B)}{P(B)}$.

We have

$$\begin{aligned} P(X = 0|X < 2) &= \frac{P\big(X = 0, X < 2\big)}{P(X < 2)} \\ &= \frac{P(X = 0)}{P(X < 2)} \\ &= \frac{P_X(0)}{P_X(0) + P_X(1)} \\ &= \frac{\frac{1}{2}}{\frac{1}{2} + \frac{1}{3}} = \frac{3}{5}. \end{aligned}$$

3. I roll two dice and observe two numbers X and Y. If $Z = X - Y$, find the range and PMF of Z.

Solution:

Note

$$R_X = R_Y = \{1, 2, 3, 4, 5, 6\}$$

and

$$P_X(k) = P_Y(k) = \begin{cases} \frac{1}{6} & \text{for } k = 1, 2, 3, 4, 5, 6 \\ 0 & \text{otherwise} \end{cases}$$

Since $Z = X - Y$, we conclude:

$R_Z = \{-5, -4, -3, -2, -1, 0, 1, 2, 3, 4, 5\}$

$$\begin{aligned} P_Z(-5) &= P(X = 1, Y = 6) \\ &= P(X = 1) \cdot P(Y = 6) \quad \text{(Since } X \text{ and } Y \text{ are independent)} \\ &= \frac{1}{6} \cdot \frac{1}{6} = \frac{1}{36} \end{aligned}$$

$$\begin{aligned} P_Z(-4) &= P(X = 1, Y = 5) + P(X = 2, Y = 6) \\ &= P(X = 1) \cdot P(Y = 5) + P(X = 2) \cdot P(Y = 6) \text{(independence)} \\ &= \frac{1}{6} \cdot \frac{1}{6} + \frac{1}{6} \cdot \frac{1}{6} = \frac{1}{18} \end{aligned}$$

Similarly:

$$\begin{aligned} P_Z(-3) &= P(X = 1, Y = 4) + P(X = 2, Y = 5) + P(X = 3, Y = 6) \\ &= P(X = 1) \cdot P(Y = 4) + P(X = 2) \cdot P(Y = 5) + \\ &\quad P(X = 3) \cdot P(Y = 6) \\ &= 3.\frac{1}{6} \cdot \frac{1}{6} = \frac{1}{12}. \end{aligned}$$

$$\begin{aligned} P_Z(-2) &= P(X = 1, Y = 3) + P(X = 2, Y = 4) + P(X = 3, Y = 5) + \\ &\quad P(X = 4, Y = 6) \\ &= P(X = 1) \cdot P(Y = 3) + P(X = 2) \cdot P(Y = 4) \\ &+ P(X = 3) \cdot P(Y = 5) + P(X = 4) \cdot P(Y = 6) \\ &= 4.\frac{1}{6} \cdot \frac{1}{6} = \frac{1}{9}. \end{aligned}$$

$$\begin{aligned}
P_Z(-1) &= P(X=1,Y=2)+P(X=2,Y=3)+P(X=3,Y=4) \\
&\quad + P(X=4,Y=5)+P(X=5,Y=6) \\
&= P(X=1)\cdot P(Y=2)+P(X=2)\cdot P(Y=3)+ \\
&\quad + P(X=3)\cdot P(Y=4)+P(X=4)\cdot P(Y=5)+ \\
&\qquad P(X=5)\cdot P(Y=6) \\
&= 5.\frac{1}{6}\cdot\frac{1}{6}=\frac{5}{36}.
\end{aligned}$$

$$\begin{aligned}
P_Z(0) &= P(X=1,Y=1)+P(X=2,Y=2)+P(X=3,Y=3) \\
&\quad + P(X=4,Y=4)+P(X=5,Y=5)+P(X=6,Y=6) \\
&= P(X=1)\cdot P(Y=1)+P(X=2)\cdot P(Y=2)+P(X=3)\cdot P(Y=3) \\
&\quad + P(X=4)\cdot P(Y=4)+P(X=5)\cdot P(Y=5)+P(X=6)\cdot P(Y=6) \\
&= 6.\frac{1}{6}\cdot\frac{1}{6}=\frac{1}{6}.
\end{aligned}$$

Note that by symmetry, we have:

$P_Z(k) = P_Z(-k)$

So,

$$\begin{cases}
P_Z(0) = \frac{1}{6} \\
P_Z(1) = P_Z(-1) = \frac{5}{36} \\
P_Z(2) = P_Z(-2) = \frac{1}{9} \\
P_Z(3) = P_Z(-3) = \frac{1}{12} \\
P_Z(4) = P_Z(-4) = \frac{1}{18} \\
P_Z(5) = P_Z(-5) = \frac{1}{36}
\end{cases}$$

5. 50 students live in a dormitory. The parking lot has the capacity for 30 cars. If each student has a car with probability $\frac{1}{2}$ (independently from other students), what is the probability that there won't be enough parking spaces for all the cars?

Solution:

If X is the number of cars owned by 50 students in the dormitory, then:

$X \sim Binomial(50, \frac{1}{2})$

Thus:

$$P(X > 30) = \sum_{k=31}^{50} \binom{50}{k} (\frac{1}{2})^k (\frac{1}{2})^{50-k}$$
$$= \sum_{k=31}^{50} \binom{50}{k} (\frac{1}{2})^{50}$$
$$= (\frac{1}{2})^{50} \sum_{k=31}^{50} \binom{50}{k}$$

7. For each of the following random variables, find $P(X > 5)$, $P(2 < X \leq 6)$ and $P(X > 5|X < 8)$. You do not need to provide the numerical values for your answers. In other words, you can leave your answers in the form of sums.

   (a) $X \sim Geometric(\frac{1}{5})$

   (b) $X \sim Binomial(10, \frac{1}{3})$

   (c) $X \sim Pascal(3, \frac{1}{2})$

   (d) $X \sim Hypergeometric(10, 10, 12)$

   (e) $X \sim Poisson(5)$

Solution:

First note that if $R_X \subset \{0, 1, 2, \cdots\}$, then

- $P(X > 5) = \sum_{k=6}^{\infty} P_X(k) = 1 - \sum_{k=0}^{5} P_X(k)$.

- $P(2 < X \leq 6) = P_X(3) + P_X(4) + P_X(5) + P_X(6)$.
- $P(X > 5|X < 8) = \frac{P(5<X<8)}{P(X<8)} = \frac{P_X(6)+P_X(7)}{\sum_{k=0}^{7} P_X(k)}$.

So,

(a) $X \sim Geometric(\frac{1}{5}) \longrightarrow \quad P_X(k) = (\frac{4}{5})^{k-1}(\frac{1}{5}) \quad$ for $k = 1, 2, 3, \cdots$
Therefore,

$$\begin{aligned}
P(X > 5) &= 1 - \sum_{k=1}^{5}(\frac{4}{5})^{k-1}(\frac{1}{5}) \\
&= 1 - (\frac{1}{5}) \cdot (1 + (\frac{4}{5}) + (\frac{4}{5})^2 + (\frac{4}{5})^3 + (\frac{4}{5})^4) \\
&= 1 - (\frac{1}{5}) \cdot \frac{1 - (\frac{4}{5})^5}{1 - (\frac{4}{5})} = (\frac{4}{5})^5.
\end{aligned}$$

Note that we can obtain this result directly from the random experiment behind the geometric random variable:

$P(X < 5) = P(\text{No heads in 5 coin tosses}) = (\frac{4}{5})^5$

$$\begin{aligned}
P(2 < X \leq 6) &= P_X(3) + P_X(4) + P_X(5) + P_X(6) \\
&= (\frac{1}{5})(\frac{4}{5})^2 + (\frac{1}{5})(\frac{4}{5})^3 + (\frac{1}{5})(\frac{4}{5})^4 + (\frac{1}{5})(\frac{4}{5})^5 \\
&= (\frac{1}{5})(\frac{4}{5})^2 \cdot (1 + \frac{4}{5} + (\frac{4}{5})^2 + (\frac{4}{5})^3) \\
&= (\frac{4}{5})^2(1 - (\frac{4}{5})^4).
\end{aligned}$$

$$\begin{aligned}
P(X > 5|X < 8) &= \frac{P(5 < X < 8)}{P(X < 8)} = \frac{P_X(6) + P_X(7)}{\sum_{k=1}^{7} P_X(k)} \\
&= \frac{(\frac{1}{5})((\frac{4}{5})^5 + (\frac{4}{5})^6)}{(\frac{1}{5})\sum_{k=1}^{7}(\frac{4}{5})^{k-1}} \\
&= \frac{(\frac{4}{5})^5 + (\frac{4}{5})^6}{1 + (\frac{4}{5}) + \cdots (\frac{4}{5})^6}
\end{aligned}$$

(b) $X \sim Binomial(10, \frac{1}{3}) \longrightarrow \quad P_X(k) = \binom{10}{k}(\frac{1}{3})^k(\frac{2}{3})^{10-k}$ for $k = 0, 1, 2, \cdots, 10$

So,

$$\begin{aligned}
P(X > 5) &= 1 - \sum_{k=0}^{5} \binom{10}{k} (\frac{1}{3})^k (\frac{2}{3})^{10-k} \\
&= 1 - [\binom{10}{0}(\frac{1}{3})^0(\frac{2}{3})^{10} + \binom{10}{1}(\frac{1}{3})^1(\frac{2}{3})^9 + \binom{10}{2}(\frac{1}{3})^2(\frac{2}{3})^8 \\
&+ \binom{10}{3}(\frac{1}{3})^3(\frac{2}{3})^7 + \binom{10}{4}(\frac{1}{3})^4(\frac{2}{3})^6 + \binom{10}{5}(\frac{1}{3})^5(\frac{2}{3})^5].
\end{aligned}$$

We can also solve this in a more direct way:

$$\begin{aligned}
P(X > 5) &= \sum_{k=6}^{10} \binom{10}{k} (\frac{1}{3})^k (\frac{2}{3})^{10-k} \\
&= \binom{10}{6}(\frac{1}{3})^6(\frac{2}{3})^4 + \binom{10}{7}(\frac{1}{3})^7(\frac{2}{3})^3 + \binom{10}{8}(\frac{1}{3})^8(\frac{2}{3})^2 \\
&+ \binom{10}{9}(\frac{1}{3})^9(\frac{2}{3})^1 + \binom{10}{10}(\frac{1}{3})^{10}(\frac{2}{3})^0 \\
&= (\frac{1}{3})^{10} \cdot (\binom{10}{6}2^4 + \binom{10}{7}2^3 + \binom{10}{8}2^2 + \binom{10}{9}2 + \binom{10}{10}) \\
&= (\frac{1}{3})^{10} \cdot (\binom{10}{6}2^4 + \binom{10}{7}2^3 + \binom{10}{8}2^2 + 21).
\end{aligned}$$

$$\begin{aligned}
P(2 < X \leq 6) &= P_X(3) + P_X(4) + P_X(5) + P_X(6) \\
&= \binom{10}{3}(\frac{1}{3})^3(\frac{2}{3})^7 + \binom{10}{4}(\frac{1}{3})^4(\frac{2}{3})^6 \\
&+ \binom{10}{5}(\frac{1}{3})^5(\frac{2}{3})^5 + \binom{10}{6}(\frac{1}{3})^6(\frac{2}{3})^4 \\
&= (\frac{1}{3})^{10}[\binom{10}{3}2^7 + \binom{10}{4}2^6 + \binom{10}{5}2^5 + \binom{10}{6}2^4] \\
&= 2^4(\frac{1}{3})^{10}[\binom{10}{3}2^3 + \binom{10}{4}2^2 + \binom{10}{5}2 + \binom{10}{6}].
\end{aligned}$$

$$
\begin{aligned}
P(X>5|X<8) &= \frac{P(5<X<8)}{P(X<8)} = \frac{P_X(6)+P_X(7)}{\sum_{k=0}^{7} P_X(k)} \\
&= \frac{P_X(6)+P_X(7)}{1-P_X(8)-P_X(9)-P_X(10)} \\
&= \frac{\binom{10}{6}(\frac{1}{3})^6(\frac{2}{3})^4+\binom{10}{7}(\frac{1}{3})^7(\frac{2}{3})^3}{1-(\binom{10}{8}(\frac{1}{3})^8(\frac{2}{3})^2+\binom{10}{9}(\frac{1}{3})^9(\frac{2}{3})^1+\binom{10}{10}(\frac{1}{3})^{10}(\frac{2}{3})^0)} \\
&= \frac{(\frac{1}{3})^{10}(2^4\binom{10}{6}+2^3\binom{10}{7})}{1-((\frac{1}{3})^{10}(2^2\binom{10}{8}+2\binom{10}{9}+\binom{10}{10}))} \\
&= \frac{(\frac{1}{3})^{10}(2^4\binom{10}{6}+2^3\binom{10}{7})}{1-((\frac{1}{3})^{10}(2^2\times 45+2\times 10+1))} \\
&= \frac{(\frac{1}{3})^{10}\times 2^3(2\binom{10}{6}+\binom{10}{7})}{1-((\frac{1}{3})^{10}\times 201)} \\
&= \frac{2^3(2\binom{10}{6}+\binom{10}{7})}{3^{10}-201}
\end{aligned}
$$

(c) $X \sim Pascal(3, \frac{1}{2}) \longrightarrow \quad P_X(k) = \binom{k-1}{2}(\frac{1}{2})^k \quad \text{for } k = 3, 4, 5, \cdots$

So:

$$
\begin{aligned}
P(X>5) &= 1-\sum_{k=3}^{5}\binom{k-1}{2}(\frac{1}{2})^k \\
&= 1-(\binom{2}{2}(\frac{1}{2})^3+\binom{3}{2}(\frac{1}{2})^4+\binom{4}{2}(\frac{1}{2})^5) \\
&= 1-((\frac{1}{2})^3+3(\frac{1}{2})^4+6(\frac{1}{2})^5) \\
&= 1-(\frac{1}{2})^5(4+6+6) \\
&= 1-((\frac{1}{2})^5\times 2^4) = \frac{1}{2}.
\end{aligned}
$$

$$\begin{aligned}
P(2 < X \leq 6) &= P_X(3) + P_X(4) + P_X(5) + P_X(6) \\
&= \binom{2}{2}(\frac{1}{2})^3 + \binom{3}{2}(\frac{1}{2})^4 + \binom{4}{2}(\frac{1}{2})^5 + \binom{5}{2}(\frac{1}{2})^6 \\
&= (\frac{1}{2})^3 + 3(\frac{1}{2})^4 + 6(\frac{1}{2})^5 + 10(\frac{1}{2})^6 \\
&= (\frac{1}{2})^6(8 + 3 \times 4 + 6 \times 2 + 10) = 42 \times (\frac{1}{2})^6 = \frac{21}{32}.
\end{aligned}$$

$$\begin{aligned}
P(X > 5|X < 8) &= \frac{P(5 < X < 8)}{P(X < 8)} = \frac{P_X(6) + P_X(7)}{\sum_{k=3}^{7} P_X(k)} \\
&= \frac{\binom{5}{2}(\frac{1}{2})^6 + \binom{6}{2}(\frac{1}{2})^7}{\binom{2}{2}(\frac{1}{2})^3 + \binom{3}{2}(\frac{1}{2})^4 + \binom{4}{2}(\frac{1}{2})^5 + \binom{5}{2}(\frac{1}{2})^6 + \binom{6}{2}(\frac{1}{2})^7} \\
&= \frac{10(\frac{1}{2})^6 + (\frac{1}{2})^7}{(\frac{1}{2})^3 + 3(\frac{1}{2})^4 + 6(\frac{1}{2})^5 + 10(\frac{1}{2})^6 + 15(\frac{1}{2})^7} \\
&= \frac{20 + 15}{16 + 24 + 24 + 20 + 15} = \frac{35}{99}.
\end{aligned}$$

(d) $X \sim Hypergeometric(10, 10, 12)$ $b = r = 10, k = 12$

$R_X = \{max(0, k - r), \cdots, min(k, b)\} = \{2, 3, 4, \cdots, 10\}$

So:

$P_X(k) = \frac{\binom{10}{k}\binom{10}{12-k}}{\binom{20}{12}}$ for $k = 2, 3, \cdots, 10$

$$\begin{aligned}
P(X > 5) &= 1 - \sum_{k=2}^{5} \frac{\binom{10}{k}\binom{10}{12-k}}{\binom{20}{12}} \\
&= 1 - \left[\frac{\binom{10}{2}\binom{10}{10}}{\binom{20}{12}} + \frac{\binom{10}{3}\binom{10}{9}}{\binom{20}{12}} + \frac{\binom{10}{4}\binom{10}{8}}{\binom{20}{12}} + \frac{\binom{10}{5}\binom{10}{7}}{\binom{20}{12}}\right] \\
&= 1 - \frac{1}{\binom{20}{12}}\left[\binom{10}{2} + 10 \cdot \binom{10}{3} + \binom{10}{4}\binom{10}{8} + \binom{10}{5}\binom{10}{7}\right]
\end{aligned}$$

$$\begin{aligned}
P(2 < X \leq 6) &= P_X(3) + P_X(4) + P_X(5) + P_X(6) \\
&= \frac{\binom{10}{3}\binom{10}{9}}{\binom{20}{12}} + \frac{\binom{10}{4}\binom{10}{8}}{\binom{20}{12}} + \frac{\binom{10}{5}\binom{10}{7}}{\binom{20}{12}} + \frac{\binom{10}{6}\binom{10}{6}}{\binom{20}{12}} \\
&= \frac{1}{\binom{20}{12}}\left[\binom{10}{3}\binom{10}{9} + \binom{10}{4}\binom{10}{8} + \binom{10}{5}\binom{10}{7} + \binom{10}{6}\binom{10}{6}\right] \\
&= \frac{1}{\binom{20}{12}}\left[10 \times \binom{10}{3} + \binom{10}{4}\binom{10}{8} + \binom{10}{5}\binom{10}{7} + \binom{10}{6}\binom{10}{6}\right].
\end{aligned}$$

$$\begin{aligned}
P(X > 5 | X < 8) &= \frac{P(5 < X < 8)}{P(X < 8)} = \frac{P_X(6) + P_X(7)}{\sum_{k=2}^{7} P_X(k)} \\
&= \frac{\frac{\binom{10}{6}\binom{10}{6}}{\binom{20}{12}} + \frac{\binom{10}{7}\binom{10}{5}}{\binom{20}{12}}}{\frac{\binom{10}{2}\binom{10}{10}}{\binom{20}{12}} + \frac{\binom{10}{3}\binom{10}{9}}{\binom{20}{12}} + \frac{\binom{10}{4}\binom{10}{8}}{\binom{20}{12}} + \frac{\binom{10}{5}\binom{10}{7}}{\binom{20}{12}} + \frac{\binom{10}{6}\binom{10}{6}}{\binom{20}{12}} + \frac{\binom{10}{7}\binom{10}{5}}{\binom{20}{12}}} \\
&= \frac{\binom{10}{6}\binom{10}{6} + \binom{10}{7}\binom{10}{5}}{\binom{10}{2}\binom{10}{10} + \binom{10}{3}\binom{10}{9} + \binom{10}{4}\binom{10}{8} + \binom{10}{5}\binom{10}{7} + \binom{10}{6}\binom{10}{6} + \binom{10}{7}\binom{10}{5}}.
\end{aligned}$$

(e) $X \sim Poisson(5)$

$P_X(k) = \frac{e^{-5}5^k}{k!}$ for $k = 0, 1, 2, \cdots$

$$\begin{aligned}
P(X > 5) &= 1 - \sum_{k=0}^{5} \frac{e^{-5}5^k}{k!} \\
&= 1 - \left(\frac{5^0 e^{-5}}{0!} + \frac{5^1 e^{-5}}{1!} + \frac{5^2 e^{-5}}{2!} + \frac{5^3 e^{-5}}{3!} + \frac{5^4 e^{-5}}{4!} + \frac{5^5 e^{-5}}{5!}\right) \\
&= 1 - \left(e^{-5} + 5e^{-5} + \frac{25e^{-5}}{2} + \frac{5^3 e^{-5}}{3!} + \frac{5^4 e^{-5}}{4!} + \frac{5^5 e^{-5}}{5!}\right) \\
&= 1 - e^{-5}\left(6 + \frac{25}{2} + \frac{5^3}{3!} + \frac{5^4}{4!} + \frac{5^5}{5!}\right).
\end{aligned}$$

$$\begin{aligned} P(2 < X \leq 6) &= P_X(3) + P_X(4) + P_X(5) + P_X(6) \\ &= \frac{e^{-5}5^3}{3!} + \frac{e^{-5}5^4}{4!} + \frac{e^{-5}5^5}{5!} + \frac{e^{-5}5^6}{6!} \\ &= e^{-5}(\frac{5^3}{3!} + \frac{5^4}{4!} + \frac{5^5}{5!} + \frac{5^6}{6!}). \end{aligned}$$

$$\begin{aligned} P(X > 5|X < 8) &= \frac{P(5 < X < 8)}{P(X < 8)} = \frac{P_X(6) + P_X(7)}{\sum_{k=0}^{7} P_X(k)} \\ &= \frac{e^{-5}(\frac{5^6}{6!} + \frac{5^7}{7!})}{e^{-5}(\frac{5^0}{0!} + \frac{5^1}{1!} + \frac{5^2}{2!} + \frac{5^3}{3!} + \frac{5^4}{4!} + \frac{5^5}{5!} + \frac{5^6}{6!} + \frac{5^7}{7!})} \\ &= \frac{\frac{5^6}{6!} + \frac{5^7}{7!}}{\frac{5^0}{0!} + \frac{5^1}{1!} + \frac{5^2}{2!} + \frac{5^3}{3!} + \frac{5^4}{4!} + \frac{5^5}{5!} + \frac{5^6}{6!} + \frac{5^7}{7!}} \\ &= \frac{\frac{5^6}{6!} + \frac{5^7}{7!}}{6 + \frac{5^2}{2!} + \frac{5^3}{3!} + \frac{5^4}{4!} + \frac{5^5}{5!} + \frac{5^6}{6!} + \frac{5^7}{7!}} \end{aligned}$$

9. In this problem, we would like to show that the geometric random variable is **memoryless**. Let $X \sim Geometric(p)$. Show that

$$P(X > m + l|X > m) = P(X > l), \qquad \text{for } m, l \in \{1, 2, 3, \cdots\}$$

We can interpret this in the following way: remember that a geometric random variable can be obtained by tossing a coin repeatedly until observing the first heads. If we toss the coin several times and do not observe a heads, from now on it is as if we start all over again. In other words, the failed coin tosses do not impact the distribution of waiting time from now on. The reason for this is that the coin tosses are independent.

Solution:

Since $X \sim Geometric(p)$, we have:

$P_X(k) = (1-p)^{k-1}p$ for $k = 1, 2, ...$

Thus:

$$\begin{aligned} P(X > m) &= \sum_{k=m+1}^{\infty} (1-p)^{k-1}p \\ &= (1-p)^m p \sum_{k=0}^{\infty} (1-p)^k \\ &= p(1-p)^m \frac{1}{1-(1-p)} \\ &= (1-p)^m. \end{aligned}$$

Similarly,

$$P(X > m + l) = (1-p)^{m+l}.$$

Therefore:

$$\begin{aligned} P(X > m + l | X > m) &= \frac{P(X > m + l \text{ and } P(X > m))}{P(X > m)} \\ &= \frac{P(X > m + l)}{P(X > m)} \\ &= \frac{(1-p)^{m+l}}{(1-p)^m} \\ &= (1-p)^l \\ &= P(X > l). \end{aligned}$$

11. The number of emails that I get in a weekday (Monday through Friday) can be modeled by a Poisson distribution with an average of $\frac{1}{6}$ emails per minute. The number of emails that I receive on weekends (Saturday and Sunday) can be modeled by a Poisson distribution with an average of $\frac{1}{30}$ emails per minute.

1. What is the probability that I get no emails in an interval of length 4 hours on a Sunday?
2. A random day is chosen (all days of the week are equally likely to be selected), and a random interval of length one hour is selected in the chosen day. It is observed that I did not receive any emails in that interval. What is the probability that the chosen day is a weekday?

Solution:

(a)

$$T = 4 \times 60 = 240 \text{ min}$$
$$\lambda = 240 \times \frac{1}{30} = 8$$

Thus $X \sim Poisson(\lambda = 8)$

$$P(X = 0) = e^{-\lambda} = e^{-8}$$

(b) Let D be the event that a weekday is chosen and let E be the event that a Saturday or Sunday is chosen.

Then:

$$P(D) = \frac{5}{7}$$
$$P(E) = \frac{2}{7}.$$

Let A be the event that I receive no emails during the chosen interval then:

$$P(A|D) = e^{-\lambda_1} = e^{-\frac{1}{6}\cdot 60} = e^{-10}$$
$$P(A|E) = e^{-\lambda_2} = e^{-\frac{1}{30}\cdot 60} = e^{-2}.$$

Therefore:

$$
\begin{aligned}
P(D|A) = \frac{P(A|D).P(D)}{P(A)} &= \frac{e^{-10}\frac{5}{7}}{P(A|D)P(D) + P(A|E)P(E)} \\
&= \frac{e^{-10}\frac{5}{7}}{e^{-10}\frac{5}{7} + e^{-2}\frac{2}{7}} \\
&= \frac{5}{5 + 2e^{8}} \approx 8.4 \times 10^{-4}.
\end{aligned}
$$

13. Let X be a discrete random variable with the following CDF:

$$
F_X(x) = \begin{cases} 0 & \text{for } x < 0 \\ \frac{1}{6} & \text{for } 0 \leq x < 1 \\ \frac{1}{2} & \text{for } 1 \leq x < 2 \\ \frac{3}{4} & \text{for } 2 \leq x < 3 \\ 1 & \text{for } x \geq 3 \end{cases}
$$

Find the range and PMF of X.

Solution:

$$R_X = \{0, 1, 2, 3\}.$$

$$P_X(x) = F_X(x) - F_X(x - \epsilon).$$

$$P_X(0) = F_X(0) - F_X(0-\epsilon) = \frac{1}{6} - 0 = \frac{1}{6}$$
$$P_X(1) = F_X(1) - F_X(1-\epsilon) = \frac{1}{2} - \frac{1}{6} = \frac{1}{3}$$
$$P_X(2) = F_X(2) - F_X(2-\epsilon) = \frac{3}{4} - \frac{1}{2} = \frac{1}{4}$$
$$P_X(3) = F_X(3) - F_X(3-\epsilon) = 1 - \frac{3}{4} = \frac{1}{4}.$$

$$P_X(x) = \begin{cases} \frac{1}{6} & \text{for } x = 0 \\ \frac{1}{3} & \text{for } x = 1 \\ \frac{1}{4} & \text{for } x = 2 \\ \frac{1}{4} & \text{for } x = 3 \\ 0 & \text{otherwise} \end{cases}$$

15. Let $X \sim Geometric(\frac{1}{3})$ and let $Y = |X - 5|$. Find the range and PMF of Y.

Solution:

$$R_X = \{1, 2, 3, ...\}$$

$$P_X(k) = \frac{1}{3}\left(\frac{2}{3}\right)^{k-1}, \qquad \text{for } k = 1, 2, 3, ...$$

Thus,

$$R_Y = \{|X-5| \big| X \in R_X\} = 0, 1, 2, \ldots.$$

Thus,

$$\begin{aligned} P_Y(0) &= P(Y=0) = P(|X-5|=0) = P(X=5) \\ &= (\frac{2}{3})^4(\frac{1}{3}). \end{aligned}$$

For $k = 1, 2, 3, 4$

$$\begin{aligned} P_Y(k) &= P(Y=k) = P(|X-5|=k) = P(X=5+k \quad \text{or} \quad X=5-k) \\ &= P_X(5+k) + P_X(5-k) = [(\frac{2}{3})^{4+k} + (\frac{2}{3})^{4-k}](\frac{1}{3}). \end{aligned}$$

For $k \geq 5$,

$$\begin{aligned} P_Y(k) &= P(Y=k) = P(|X-5|=k) = P(X=5+k) \\ &= P_X(5+k) = (\frac{2}{3})^{4+k}(\frac{1}{3}). \end{aligned}$$

So, in summary:

$$P_Y(k) = \begin{cases} (\frac{2}{3})^{k+4}(\frac{1}{3}) & \text{for } k = 0, 5, 6, 7, 8, \ldots \\ \\ ((\frac{2}{3})^{k+4} + (\frac{2}{3})^{4-k})(\frac{1}{3}) & \text{for } k = 1, 2, 3, 4 \\ \\ 0 & \text{otherwise} \end{cases}$$

17. Let $X \sim Geometric(p)$. Find $\mathrm{Var}(X)$.

Solution: First, note:

$$\sum_{k=0}^{\infty} x^k = \frac{1}{1-x} \quad \text{for } |x| < 1.$$

Taking the derivative:

$$\sum_{k=1}^{\infty} kx^{k-1} = \frac{1}{(1-x)^2} \quad \text{for } |x| < 1.$$

Taking another derivative:

$$\sum_{k=2}^{\infty} k(k-1)x^{k-2} = \frac{2}{(1-x)^3} \quad \text{for } |x| < 1.$$

Now we can use the above identities to find $\mathrm{Var}(X)$. If $X \sim Geometric(p)$, then

$$P_X(k) = p(1-p)^{k-1} = pq^{k-1} \text{ for } k = 1, 2, ...$$

where $q = 1 - p$. Thus

$$\begin{aligned} EX &= p \sum_{k=1}^{\infty} kq^{k-1} \\ &= p \frac{1}{(1-q)^2} = \frac{1}{p}. \end{aligned}$$

$$\begin{aligned}
E[X(X-1)] &= p\sum_{k=1}^{\infty} k(k-1)q^{k-1} \quad \text{by LOTUS} \\
&= pq\sum_{k=2}^{\infty} k(k-1)q^{k-2} = pq\frac{2}{(1-q)^3} \\
&= \frac{2pq}{p^3} = \frac{2q}{p^2}.
\end{aligned}$$

Thus:

$$\begin{aligned}
EX^2 - EX &= \frac{2q}{p^2} \\
EX^2 &= \frac{2q}{p^2} + \frac{1}{p}.
\end{aligned}$$

Therefore:

$$\begin{aligned}
\text{Var}(X) = EX^2 - (EX)^2 &= \frac{2q}{p^2} + \frac{1}{p} - \frac{1}{p^2} \\
&= \frac{2(1-p)+p-1}{p^2} = \frac{1-p}{p^2}.
\end{aligned}$$

19. Suppose that $Y = -2X + 3$. If we know $EY = 1$ and $EY^2 = 9$, find EX and $\text{Var}(X)$.

Solution:

$$Y = -2X + 3$$

$$EY = -2EX + 3 \quad \text{linearity of expectation}$$

$$1 = -2EX + 3 \quad \rightarrow \qquad\qquad EX = 1$$

$$\text{Var}(Y) = 4 \times \text{Var}(X) = EY^2 - (EY)^2 = 9 - 1 = 8$$
$$\rightarrow \quad \text{Var}(X) = 2$$

21. (Coupon collector's problem) Suppose that there are N different types of coupons. Each time you get a coupon, it is equally likely to be any of the N possible types. Let X be the number of coupons you will need to get before having observed each coupon at least once.

    (a) Show that you can write $X = X_0 + X_1 + \cdots + X_{N-1}$, where $X_i \sim Geometric(\frac{N-i}{N})$.
    (b) Find EX.

    Solution:

    (a) After you have already collected i distinct coupons, define X_i to be the number of additional coupons you need to collect in order to get the $i+1$'th distinct coupon. Then, we have $X_0 = 1$, since the first coupon you collect is always a new one. Then, X_1 will be a geometric random variable with success probability of $p_2 = \frac{N-1}{N}$. More generally, we can write $X_i \sim Geometric(\frac{N-i}{N})$, for $i = 0, 1, ..., N-1$. Note that by definition write $X = X_0 + X_2 + \cdots + X_{N-1}$.

    (b) By linearity of expectation, we have

$$\begin{aligned} EX &= EX_0 + EX_1 + \cdots + EX_{N-1} \\ &= 1 + \frac{N}{N-1} + \frac{N}{N-2} + \cdots + \frac{N}{1} \\ &= N\left[1 + \frac{1}{2} + \cdots + \frac{1}{N-1} + \frac{1}{N}\right] \end{aligned}$$

23. Let X be a random variable with mean $EX = \mu$. Define the function $f(\alpha)$ as

$$f(\alpha) = E[(X - \alpha)^2].$$

Find the value of α that minimizes f.

Solution:

$$\begin{aligned} f(\alpha) &= E(X^2 - 2\alpha X + \alpha^2) \\ &= EX^2 - 2\alpha EX + \alpha^2. \end{aligned}$$

Thus:

$$f(\alpha) = \alpha^2 - 2(EX)\alpha + EX^2.$$

$f(\alpha)$ is a polynomial of degree 2 with positive coefficient for α^2

$$\begin{aligned} \frac{\partial f(\alpha)}{\partial \alpha} = 0 \quad &\rightarrow \quad 2\alpha - 2EX = 0 \\ &\rightarrow \quad \alpha = EX \end{aligned}$$

25. The **median** of a random variable X is defined as any number m that satisfies both of the following conditions:

$$P(X \geq m) \geq \frac{1}{2} \qquad \text{and} \qquad P(X \leq m) \geq \frac{1}{2}.$$

Note that the median of X is not necessarily unique. Find the median of X if

(a) The PMF of X is given by

$$P_X(k) = \begin{cases} 0.4 & \text{for } k = 1 \\ 0.3 & \text{for } k = 2 \\ 0.3 & \text{for } k = 3 \\ 0 & \text{otherwise} \end{cases}$$

(b) X is the result of a rolling of a fair die.

(c) $X \sim Geometric(p)$, where $0 < p < 1$.

Solution: (a) $m = 2$, since

$$P(X \geq 2) = 0.6 \text{ and } P(X \leq 2) = 0.7$$

(b)

$$\begin{aligned} P_X(k) &= \frac{1}{6} \text{ for } k = 1, 2, 3, 4, 5, 6 \\ &\rightarrow 3 \leq m \leq 4 \end{aligned}$$

Thus, we conclude $3 \leq m \leq 4$. Any value $\in [3, 4]$ is a median for X.

(c)

$$\begin{aligned} P_X(k) &= (1-p)^{k-1}p = q^{k-1}p \quad \text{ where } q = 1 - p \\ P(X \leq m) &= \sum_{k=1}^{\lfloor m \rfloor} q^{k-1}p = p(1 + q + \cdots q^{m-1}) \\ &= p\frac{1 - q^{\lfloor m \rfloor}}{1 - q} = 1 - q^{\lfloor m \rfloor}, \end{aligned}$$

where $\lfloor m \rfloor$ is the largest integer less than or equal to m. We need $1 - q^{\lfloor m \rfloor} \geq \frac{1}{2}$.

Therefore:

$$q^{\lfloor m \rfloor} \leq \frac{1}{2} \quad \to \lfloor m \rfloor \log_2(q) \leq -1 \quad \to \lfloor m \rfloor \log_2 \frac{1}{q} \geq 1$$
$$\to \lfloor m \rfloor \geq \frac{1}{\log_2 \frac{1}{q}}$$

Also

$$\begin{aligned} P(X \geq m) &= \sum_{k=\lceil m \rceil}^{\infty} q^{k-1} p = pq^{\lceil m \rceil - 1}(1 + q + \cdots) \\ &= p\frac{q^{\lceil m \rceil} - 1}{1-q} = q^{\lceil m \rceil - 1}, \end{aligned}$$

where $\lceil m \rceil$ is the smallest integer larger than or equal to m. Thus:

$$\begin{aligned} q^{\lceil m \rceil - 1} \geq \frac{1}{2} \quad &\to (\lceil m \rceil - 1) \log_2 q \geq -1 \\ &\to (\lceil m \rceil - 1) \log_2 (\frac{1}{q}) \leq 1 \quad \to \lceil m \rceil - 1 \leq \frac{1}{\log_2(\frac{1}{q})} \\ &\to \lceil m \rceil \leq \frac{1}{\log_2(\frac{1}{q})} + 1 \end{aligned}$$

Thus any m satisfying

$$\lfloor m \rfloor \geq \frac{1}{\log_2 \frac{1}{q}} \text{ and } \lceil m \rceil \leq \frac{1}{\log_2(\frac{1}{q})} + 1$$

is a median for X. For example if $p = \frac{1}{5}$ then $\lfloor m \rfloor \geq 3.1$ and $\lceil m \rceil \leq 4.1$. So $m = 4$.

Chapter 4

Continuous and Mixed Random Variables

1. I choose a real number uniformly at random in the interval $[2, 6]$ and call it X.

   (a) Find the CDF of X, $F_X(x)$.

   (b) Find EX.

Solution:
(a) We saw that all individual points have probability 0; i.e.,$P(X = x) = 0$ for all x in uniform distribution. Also, the uniformity implies that the probability of an interval of length l in $[a, b]$ must be proportional to its length:

$$P(X \in [x_1, x_2]) \propto (x_2 - x_1), \qquad \text{where } 2 \leq x_1 \leq x_2 \leq 6.$$

Since $P(X \in [2, 6]) = 1$, we conclude

$$P(X \in [x_1, x_2]) = \frac{x_2 - x_1}{6 - 2} = \frac{x_2 - x_1}{4}, \qquad \text{where } 2 \leq x_1 \leq x_2 \leq 6.$$

Now, let us find the CDF. By definition $F_X(x) = P(X \leq x)$, thus we immediately have

$$
\begin{aligned}
F_X(x) &= 0, \qquad \text{for } x < 2, \\
F_X(x) &= 1, \qquad \text{for } x \geq 6.
\end{aligned}
$$

For $2 \leq x \leq 6$, we have

$$
\begin{aligned}
F_X(x) &= P(X \leq x) \\
&= P(X \in [2, x]) \\
&= \frac{x-2}{4}.
\end{aligned}
$$

Thus, to summarize

$$
F_X(x) = \begin{cases} 0 & \text{for } x < 2 \\ \frac{x-2}{4} & \text{for } 2 \leq x \leq 6 \\ 1 & \text{for } x > 6 \end{cases}
$$

(b) As we saw, the PDF of X is given by

$$
f_X(x) = \begin{cases} \frac{1}{6-2} = \frac{1}{4} & 2 \leq x \leq 6 \\ 0 & x < 2 \text{ or } x > 6. \end{cases}
$$

So, to find its expected value, we can write

$$
\begin{aligned}
EX &= \int_{-\infty}^{\infty} x f_X(x) dx \\
&= \int_2^6 x(\frac{1}{4}) dx \\
&= \frac{1}{4}\left[\frac{1}{2}x^2\right]_2^6 = 4.
\end{aligned}
$$

Note: An easier way to derive the CDF of X and EX is to use the relations for uniform distributions:

As we saw, if $X \sim Uniform(a, b)$ then the CDF and expected value of X are given by

$$F_X(x) = \begin{cases} 0 & x < a \\ \frac{x-a}{b-a} & a \leq x \leq b \\ 1 & x > b \end{cases}$$

$$EX = \frac{a+b}{2}$$

So, we could also directly write $F_X(x)$ and EX using the above formulas and get the same results.

3. Let X be a continuous random variable with PDF

$$f_X(x) = \begin{cases} x^2 + \frac{2}{3} & 0 \leq x \leq 1 \\ 0 & \text{otherwise} \end{cases}$$

   (a) Find $E(X^n)$, for $n = 1, 2, 3, \cdots$.

   (b) Find variance of X.

Solution:

(a) Using LOTUS, we have

$$\begin{aligned} E[X^n] &= \int_{-\infty}^{\infty} x^n f_X(x) dx \\ &= \int_0^1 x^n (x^2 + \frac{2}{3}) dx \\ &= \int_0^1 (x^{n+2} + \frac{2}{3} x^n) dx \\ &= \left[\frac{1}{n+3} x^{n+3} + \frac{2}{3(n+1)} x^{n+1} \right]_0^1 \\ &= \frac{1}{n+3} + \frac{2}{3(n+1)} \\ &= \frac{5n+9}{3(n+1)(n+3)}. \quad \text{where } n = 1, 2, 3, \cdots \end{aligned}$$

(b) We know that

$$\text{Var}(X) = EX^2 - (EX)^2.$$

So, we need to find the values of EX and EX^2

$$E[X] = \frac{7}{12}$$

$$E[X^2] = \frac{19}{45}$$

Thus, we have

$$\text{Var}(X) = EX^2 - (EX)^2 = \frac{19}{45} - (\frac{7}{12})^2 = 0.0819.$$

5. Let X be a continuous random variable with PDF

$$f_X(x) = \begin{cases} \frac{5}{32}x^4 & 0 \leq x \leq 2 \\ 0 & \text{otherwise} \end{cases}$$

and let $Y = X^2$.

(a) Find CDF of Y.

(b) Find PDF of Y.

(c) Find EY.

Solution:

(a) First, we note that $R_Y = [0, 4]$. As usual, we start with the CDF. For $y \in [0, 4]$, we have

$$\begin{aligned}
F_Y(y) &= P(Y \leq y) \\
&= P(X^2 \leq y) \\
&= P(0 \leq X \leq \sqrt{y}) \quad \text{since } x \text{ is not negative} \\
&= \int_0^{\sqrt{y}} \frac{5}{32} x^4 dx \\
&= \frac{1}{32} (\sqrt{y})^5 \\
&= \frac{1}{32} y^2 \sqrt{y}
\end{aligned}$$

Thus, the CDF of Y is given by

$$F_Y(y) = \begin{cases} 0 & \text{for } y < 0 \\ \frac{1}{32} y^2 \sqrt{y} & \text{for } 0 \leq y \leq 4 \\ 1 & \text{for } y > 4. \end{cases}$$

(b)

$$f_Y(y) = \frac{d}{dy} F_Y(y) = \begin{cases} \frac{5}{64} y \sqrt{y} & \text{for } 0 \leq y \leq 4 \\ 0 & \text{otherwise} \end{cases}$$

(c) To find the EY, we can directly apply LOTUS,

$$\begin{aligned}
E[Y] = E[X^2] &= \int_{-\infty}^{\infty} x^2 f_X(x) dx \\
&= \int_0^2 x^2 \cdot \frac{5}{32} x^4 dx \\
&= \int_0^2 \frac{5}{32} x^6 dx \\
&= \frac{5}{32} \times \frac{1}{7} \times 2^7 = \frac{20}{7}.
\end{aligned}$$

7. Let $X \sim Exponential(\lambda)$. Show that

1. $EX^n = \frac{n}{\lambda} EX^{n-1}$, for $n = 1, 2, 3, \cdots$.
2. $EX^n = \frac{n!}{\lambda^n}$.

Solution:

(a) We use integration by part (choosing $u = x^n$ and $v = -e^{-\lambda x}$)

$$\begin{aligned}
EX^n &= \int_0^\infty x^n \lambda e^{-\lambda x} \; dx \\
&= \left[-x^n e^{-\lambda x}\right]_0^\infty + n \int_0^\infty x^{n-1} e^{-\lambda x} \; dx \\
&= 0 + \frac{n}{\lambda} \int_0^\infty x^{n-1} \lambda e^{-\lambda x} \; dx \\
&= \frac{n}{\lambda} EX^{n-1}.
\end{aligned}$$

(b) We can prove this by induction using part (a). Note that for $n = 1$, we have

$$EX = \frac{1}{\lambda} = \frac{1!}{\lambda^1}.$$

Now, if we have $EX^n = \frac{n!}{\lambda^n}$, we can write

$$\begin{aligned}
EX^{n+1} &= \frac{n+1}{\lambda} EX^n \\
&= \frac{n+1}{\lambda} \cdot \frac{n!}{\lambda^n} \\
&= \frac{(n+1)!}{\lambda^{n+1}}.
\end{aligned}$$

9. Let $X \sim N(3, 9)$ and $Y = 5 - X$.

(a) Find $P(X > 2)$.

(b) Find $P(-1 < Y < 3)$.

(c) Find $P(X > 4|Y < 2)$.

Solution:

(a) Find $P(X > 2)$: We have $\mu_X = 3$ and $\sigma_X = 3$. Thus,

$$\begin{aligned} P(X > 2) &= 1 - \Phi\left(\frac{2-3}{3}\right) \\ &= 1 - \Phi\left(\frac{-1}{3}\right) = \Phi\left(\frac{1}{3}\right) \end{aligned}$$

(b) Find $P(-1 < Y < 3)$: Since $Y = 5 - X$, we have $Y \sim N(2, 9)$. Therefore,

$$\begin{aligned} P(-1 < Y < 3) &= \Phi\left(\frac{3-2}{3}\right) - \Phi\left(\frac{(-1)-2}{3}\right) \\ &= \Phi\left(\frac{1}{3}\right) - \Phi(-1). \end{aligned}$$

Note that we can also solve this in the following way:

$$\begin{aligned} P(-1 < Y < 3) &= P(-1 < 5 - X < 3) \\ &= P(2 < X < 6) \\ &= \Phi\left(\frac{6-3}{3}\right) - \Phi\left(\frac{2-3}{3}\right) \\ &= \Phi(1) - \Phi\left(-\frac{1}{3}\right) \\ &= \Phi\left(\frac{1}{3}\right) - \Phi(-1). \end{aligned}$$

(c) Find $P(X > 4|Y < 2)$:

$$\begin{aligned}
P(X > 4|Y < 2) &= P(X > 4|5 - X < 2) \\
&= P(X > 4|X > 3) \\
&= \frac{P(X > 4, X > 3)}{P(X > 3)} \\
&= \frac{P(X > 4)}{P(X > 3)} \\
&= \frac{1 - \Phi(\frac{4-3}{3})}{1 - \Phi(\frac{3-3}{3})} \\
&= \frac{1 - \Phi(\frac{1}{3})}{1 - \Phi(0)} \\
&= 2(1 - \Phi(\frac{1}{3}))
\end{aligned}$$

11. Let $X \sim Exponential(2)$ and $Y = 2 + 3X$.

    (a) Find $P(X > 2)$.

    (b) Find EY and variance of Y.

    (c) Find $P(X > 2|Y < 11)$.

Solution:

(a) Find $P(X > 2)$:

$$\begin{aligned}
P(X > 2) &= 1 - P(X \leq 2) \\
&= 1 - F_X(2) = 1 - (1 - e^{-4}) = e^{-4}
\end{aligned}$$

(b) Find EY:

·Since $Y = 2 + 3X$,

we have $EY = 2 + 3EX = 2 + 3 \times \frac{1}{2} = \frac{7}{2}$.

$\text{Var}(Y) = \text{Var}(2 + 3X) = 9 \times \text{Var}(X) = 9 \times \frac{1}{4} = \frac{9}{4}$

(c) Find $P(X > 2|Y < 11)$:

$$\begin{aligned} P(X > 2|Y < 11) &= P(X > 2|2 + 3X < 11) \\ &= P(X > 2|X < 3) \\ &= \frac{P(X > 2, X < 3)}{P(X < 3)} \\ &= \frac{P(2 < X < 3)}{P(X < 3)} \\ &= \frac{e^{-4} - e^{-6}}{1 - e^{-6}} \end{aligned}$$

13. Let X be a random variable with the following CDF:

$$F_X(x) = \begin{cases} 0 & \text{for } x < 0 \\ x & \text{for } 0 \le x < \frac{1}{4} \\ x + \frac{1}{2} & \text{for } \frac{1}{4} \le x < \frac{1}{2} \\ 1 & \text{for } x \ge \frac{1}{2} \end{cases}$$

(a) Plot $F_X(x)$ and explain why X is a mixed random variable.

(b) Find $P(X \le \frac{1}{3})$.

(c) Find $P(X \ge \frac{1}{4})$.

(d) Write the CDF of X in the form of

$$F_X(x) = C(x) + D(x),$$

where $C(x)$ is a continuous function and $D(x)$ is in the form of a staircase function, i.e.,

$$D(x) = \sum_k a_k u(x - x_k)$$

(e) Find $c(x) = \frac{d}{dx}C(x)$.

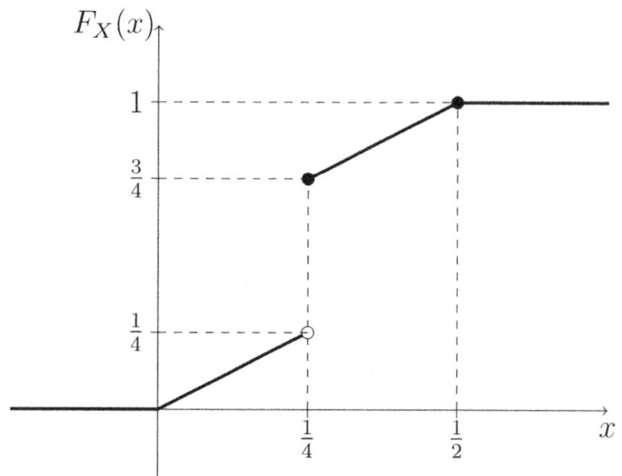

Figure 4.1: CDF of the Mixed random variable

(f) Find EX using $EX = \int_{-\infty}^{\infty} xc(x)dx + \sum_k x_k a_k$

Solution:

(a) X is a mixed random variable because the CDF is not a continuous function nor in the form of a staircase function.

(b)

$$P(X \leq \frac{1}{3}) = F_X(\frac{1}{3}) = \frac{1}{3} + \frac{1}{2} = \frac{5}{6}$$

(c)

$$\begin{aligned}
P(X \geq \frac{1}{4}) &= 1 - P(X < \frac{1}{4}) \\
&= 1 - P(X \leq \frac{1}{4}) + P(X = \frac{1}{4}) \\
&= 1 - F_X(\frac{1}{4}) + \frac{1}{2} = 1 - \frac{3}{4} + \frac{1}{2} = \frac{3}{4}
\end{aligned}$$

(d) We can write:

$F_X(x) = C(x) + D(x)$

where

$$C(x) = \begin{cases} 0 & \text{for } x < 0 \\ x & \text{for } 0 \le x \le \frac{1}{2} \\ \frac{1}{2} & \text{for } x \ge \frac{1}{2} \end{cases}$$

and

$$D(x) = \begin{cases} 0 & \text{for } x < \frac{1}{4} \\ \frac{1}{2} & \text{for } x \ge \frac{1}{4} \end{cases}$$

Thus $D(x) = \frac{1}{2}u(x - \frac{1}{4})$.

(e)

$$c(x) = \begin{cases} 0 & \text{for } x < 0 \quad \text{or} \quad x \ge \frac{1}{2} \\ 1 & \text{for } 0 \le x < \frac{1}{2} \end{cases}$$

(f)

$EX = \int_{-\infty}^{\infty} xc(x)dx + \sum_k x_k a_k = \int_0^{\frac{1}{2}} xdx + \frac{1}{2} \cdot \frac{1}{4} = \frac{1}{8} + \frac{1}{8} = \frac{1}{4}$

15. Let X be a mixed random variable with the following generalized PDF:

$$f_X(x) = \frac{1}{3}\delta(x+2) + \frac{1}{6}\delta(x-1) + \frac{1}{2}.\frac{1}{\sqrt{2\pi}}e^{-\frac{x^2}{2}}$$

(a) Find $P(X = 1)$ and $P(X = -2)$.

(b) Find $P(X \ge 1)$.

(c) Find $P(X = 1|X \geq 1)$.

(d) Find EX and $\text{Var}(X)$.

Solution:

Note that $\frac{1}{\sqrt{2\pi}}e^{-\frac{x^2}{2}}$ is the PDF of a standard normal random variable. So, we can plot the PDF of X as follows:

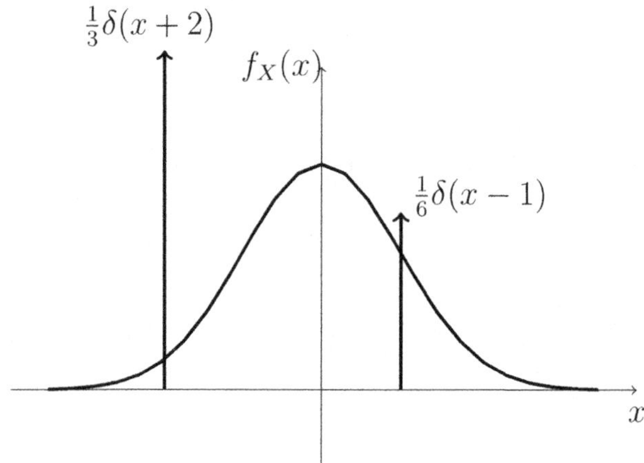

(a)

$$P(X = 1) = \frac{1}{6} \qquad P(X = -2) = \frac{1}{3}$$

(b)

$$\begin{aligned}
P(X \geq 1) &= P(X = 1) + \int_1^{\infty} \frac{1}{2}\frac{1}{\sqrt{2\pi}}e^{-\frac{x^2}{2}}\,dx \\
&= \frac{1}{6} + \frac{1}{2}[1 - \phi(\frac{1-0}{1})] \\
&= \frac{1}{6} + \frac{1}{2}[1 - \phi(1)] \\
&= \frac{1}{6} + \frac{1}{2}\phi(-1)
\end{aligned}$$

(c)

$$P(X=1|X\geq 1)=\frac{P(X=1 \text{ and } X\geq 1)}{P(X\geq 1)}$$
$$=\frac{P(X=1)}{P(X\geq 1)}=\frac{\frac{1}{6}}{\frac{1}{6}+\frac{1}{2}\phi(-1)}$$

(d)

$$EX=\frac{1}{6}\cdot 1+\frac{1}{3}\cdot(-2)+\frac{1}{2}EZ \quad \text{where } Z\sim N(0,1)$$

Thus,

$$EX=\frac{1}{6}-\frac{2}{3}+0=-\frac{1}{2}$$

$$EX^2=\int_{-\infty}^{\infty}x^2 f_X(x)dx$$
$$=\int_{-\infty}^{\infty}\left(\frac{1}{3}x^2\delta(x+2)+\frac{1}{6}x^2\delta(x-1)+\frac{1}{2}.\frac{1}{\sqrt{2\pi}}x^2e^{-\frac{x^2}{2}}\right)dx$$
$$=\frac{1}{3}\cdot(-2)^2+\frac{1}{6}\cdot 1^2+\frac{1}{2}EZ^2 \quad \text{where } Z\sim N(0,1)$$
$$=\frac{4}{3}+\frac{1}{6}+\frac{1}{2}=2$$

$$\text{Var}(X)=EX^2-(EX)^2$$
$$=2-\left(\frac{1}{2}\right)^2$$
$$=\frac{7}{4}$$

17. A continuous random variable is said to have a *Laplace*(μ, b) distribution if its PDF is given by

$$\begin{aligned} f_X(x) &= \frac{1}{2b}\exp\left(-\frac{|x-\mu|}{b}\right) \\ &= \begin{cases} \frac{1}{2b}\exp\left(\frac{x-\mu}{b}\right) & \text{if } x < \mu \\ \frac{1}{2b}\exp\left(-\frac{x-\mu}{b}\right) & \text{if } x \geq \mu \end{cases} \end{aligned}$$

where $\mu \in \mathbb{R}$ and $b > 0$.

(a) If $X \sim Laplace(0, 1)$, find EX and $\text{Var}(X)$.

(b) If $X \sim Laplace(0, 1)$ and $Y = bX + \mu$, show that $Y \sim Laplace(\mu, b)$.

(c) Let $Y \sim Laplace(\mu, b)$, where $\mu \in \mathbb{R}$ and $b > 0$. Find EY and $\text{Var}(Y)$.

Solution:

(a) $X \sim Laplace(0, 1)$, so:

$$f_X(x) = \frac{1}{2}e^{-|x|} = \begin{cases} \frac{1}{2}e^{x} & \text{for } x < 0 \\ \frac{1}{2}e^{-x} & \text{for } x \geq 0 \end{cases}$$

Since the PDF of X is symmetric around 0, we conclude $EX = 0$. More specifically,

$$\begin{aligned} EX &= \int_{-\infty}^{\infty} x f_X(x)dx = \frac{1}{2}\int_{-\infty}^{0} xe^{x}dx + \frac{1}{2}\int_{0}^{\infty} xe^{-x}dx \\ &= -\frac{1}{2}\int_{0}^{\infty} ye^{-y}dy + \frac{1}{2}\int_{0}^{\infty} xe^{-x}dx = 0 \quad (\text{let } y = -x) \end{aligned}$$

$$\begin{aligned} \text{Var}(X) = EX^2 - (EX)^2 = EX^2 &= \int_{-\infty}^{\infty} x^2 f_X(x)dx \\ = \frac{1}{2}\int_{-\infty}^{\infty} x^2 e^{-|x|}dx &= \int_{0}^{\infty} x^2 e^{-x}dx = 2 \end{aligned}$$

Another way to obtain $\text{Var}(X)$ is as follows: Note that you can interpret X in the following way. Let $W \sim Exponential(1)$. You toss a fair coin. If you observe heads, $X = W$. Otherwise, $X = -W$. Using this construction, we have $X^2 = W^2$, thus $EX^2 = EW^2 = 2$, and since $EX = 0$, we conclude that $\text{Var}(X) = 2$.

(b) $Y = g(X)$ where $g(X) = bX + \mu$, $g'(X) = b$. Thus, using the method of transformation, we can write

$$f_Y(y) = \frac{f_X(\frac{y-\mu}{b})}{b} = \frac{1}{2b}\exp(-|\frac{y-\mu}{b}|)$$

Thus: $Y \sim Laplace(\mu, b)$.

You can also show this by starting from the CDF:

$$\begin{aligned} F_Y(y) &= P(Y \leq y) \\ &= P(bX + \mu \leq y) \\ &= P(X \leq \frac{y-\mu}{b}) \\ &= F_X(\frac{y-\mu}{b}). \end{aligned}$$

Thus

$$\begin{aligned} f_Y(y) &= \frac{d}{dy}F_Y(y) \\ &= \frac{f_X(\frac{y-\mu}{b})}{b} = \frac{1}{2b}\exp(-|\frac{y-\mu}{b}|). \end{aligned}$$

(c) We can write $Y = bX + \mu$, where $X \sim Laplace(0, 1)$
Thus by part (a), $EX = 0$ and $\text{Var}(X) = 2$

$$\begin{aligned} EY &= bEX + \mu = \mu \\ \text{Var}(Y) &= b^2\text{Var}(X) = 2b^2 \end{aligned}$$

19. A continuous random variable is said to have a **standard Cauchy distribution** if its PDF is given by

$$f_X(x) = \frac{1}{\pi(1+x^2)}.$$

If X has a standard Cauchy distribution, show that EX is not well-defined. Also, show $EX^2 = \infty$.

Solution:

$$EX = \int_{-\infty}^{\infty} x f_X(x)dx = \int_{-\infty}^{\infty} \frac{x}{\pi(1+x^2)}dx$$

But, note that:

$\int_{-\infty}^{0} \frac{x}{\pi(1+x^2)}dx = -\infty$ and $\int_{0}^{\infty} \frac{x}{\pi(1+x^2)}dx = \infty$

(Note that $\int_{0}^{\infty} \frac{x}{\pi(1+x^2)}dx = \frac{1}{2\pi}\ln(1+x^2)\big|_0^{\infty} = \infty$ and $\int_{\infty}^{0} \frac{x}{\pi(1+x^2)}dx = \frac{1}{2\pi}\ln(1+x^2)\big|_{-\infty}^{0} = -\infty$)

Thus, EX is not well defined.

$$\begin{aligned}
EX^2 &= \int_{-\infty}^{\infty} \frac{x^2}{\pi(1+x^2)}dx \\
&= \int_{-\infty}^{0} \frac{x^2}{\pi(1+x^2)}dx + \int_{0}^{\infty} \frac{x^2}{\pi(1+x^2)}dx \\
&= 2\int_{0}^{\infty} \frac{x^2}{\pi(1+x^2)}dx \\
&= 2\big[x - \arctan(x)\big]_0^{\infty} = \infty.
\end{aligned}$$

21. A continuous random variable is said to have a *Pareto*(x_m, α) distribution if its PDF is given by

$$f_X(x) = \begin{cases} \alpha \dfrac{x_{\mathrm{m}}^{\alpha}}{x^{\alpha+1}} & \text{for } x \geq x_{\mathrm{m}}, \\ \\ 0 & \text{for } x < x_{\mathrm{m}}. \end{cases}$$

where $x_m, \alpha > 0$. Let $X \sim Pareto(x_m, \alpha)$.

(a) Find the CDF of X, $F_X(x)$.

(b) Find $P(X > 3x_m | X > 2x_m)$.

(c) If $\alpha > 2$, find EX and $\mathrm{Var}(X)$.

Solution:

(a)

$$f_X(x) = \begin{cases} \alpha \dfrac{x_{\mathrm{m}}^{\alpha}}{x^{\alpha+1}} & \text{for } x \geq x_{\mathrm{m}}, \\ \\ 0 & \text{for } x < x_{\mathrm{m}}. \end{cases}$$

Note that $R_X = [x_m, \infty)$,
Thus, $F_X(x) = 0$ for $x < x_m$
For $x \geq x_m$:

$$\begin{aligned} F_X(x) &= \int_{x_m}^{x} \alpha \frac{x_m^{\alpha}}{x^{\alpha+1}} dx \\ &= \Big[- \frac{x_m^{\alpha}}{x^{\alpha}} \Big]_{x_m}^{x} = 1 - \big(\frac{x_m}{x}\big)^{\alpha} \end{aligned}$$

Thus:

$$F_X(x) = \begin{cases} 1 - \left(\frac{x_m}{x}\right)^{\alpha} & \text{for } x \geq x_m \\ \\ 0 & \text{otherwise} \end{cases}$$

(b)

$$P(X > 3x_m | X > 2x_m) = \frac{P(X > 3x_m \text{ and } X > 2x_m)}{P(X > 2x_m)}$$
$$= \frac{P(X > 3x_m)}{P(X > 2x_m)} = \frac{\left(\frac{x_m}{3x_m}\right)^\alpha}{\left(\frac{x_m}{2x_m}\right)^\alpha} = \left(\frac{2}{3}\right)^\alpha$$

(c)

$$\begin{aligned} EX &= \int_{x_m}^{\infty} x \cdot \alpha \frac{x_m^\alpha}{x^{\alpha+1}} dx \\ &= \alpha x_m^\alpha \int_{x_m}^{\infty} \frac{1}{x^\alpha} dx \\ &= \alpha x_m^\alpha \frac{{x_m}^{(1-\alpha)}}{\alpha - 1} \qquad \text{since } \alpha > 1 \\ &= \frac{\alpha x_m}{\alpha - 1} \end{aligned}$$

$$\begin{aligned} EX^2 &= \int_{x_m}^{\infty} x^2 \cdot \alpha \frac{x_m^\alpha}{x^{\alpha+1}} dx \\ &= \alpha x_m^\alpha \int_{x_m}^{\infty} x^{-\alpha+1} dx \\ &= \alpha x_m^\alpha \left[\frac{1}{-\alpha + 2} x^{-\alpha+2}\right]_{x_m}^{\infty} \qquad \text{since } \alpha > 2 \\ &= \alpha x_m^\alpha \frac{x_m^{2-\alpha}}{\alpha - 2} = \frac{\alpha}{\alpha - 2} x_m^2 \qquad \text{since } \alpha > 2 \end{aligned}$$

Thus:

$$\text{Var}(X) = EX^2 - (EX)^2 = \frac{\alpha}{\alpha - 2} x_m^2 - \left(\frac{\alpha}{\alpha - 1} x_m\right)^2 = \frac{\alpha x_m^2}{(\alpha - 2)(\alpha - 1)^2}$$

23. Let X_1, X_2, $\cdots$, X_n be independent random variables with $X_i \sim Exponential(\lambda)$. Define

$$Y = X_1 + X_2 + \cdots + X_n.$$

As we will see later, Y has a **Gamma distribution** with parameters n and λ, i.e., $Y \sim Gamma(n, \lambda)$. Using this, show that if $Y \sim Gamma(n, \lambda)$, then $EY = \frac{n}{\lambda}$ and $\mathrm{Var}(Y) = \frac{n}{\lambda^2}$.

Solution:

$$Y = X_1 + X_2 + \cdots + X_n.$$

where $X_i \sim Exponential(\lambda)$

Thus:

$$\begin{aligned} EY &= EX_1 + EX_2 + \cdots + EX_n \\ &= \frac{1}{\lambda} + \frac{1}{\lambda} + \cdots + \frac{1}{\lambda} \quad \text{since } X_i \sim Exponential(\lambda) \\ &= \frac{n}{\lambda} \end{aligned}$$

$$\begin{aligned} \mathrm{Var}(Y) &= \mathrm{Var}(X_1) + \mathrm{Var}(X_2) + \cdots + \mathrm{Var}(X_n) \quad \text{since } X_i\text{'s are independent} \\ &= \frac{1}{\lambda^2} + \frac{1}{\lambda^2} + \cdots + \frac{1}{\lambda^2} \\ &= \frac{n}{\lambda^2} \end{aligned}$$

Chapter 5

Joint Distributions: Two Random Variables

1. Consider two random variables X and Y with joint PMF, given in Table 5.1.

Table 5.1: Joint PMF of X and Y in Problem 5

	$Y = 1$	$Y = 2$
$X = 1$	$\frac{1}{3}$	$\frac{1}{12}$
$X = 2$	$\frac{1}{6}$	0
$X = 4$	$\frac{1}{12}$	$\frac{1}{3}$

(a) Find $P(X \leq 2, Y > 1)$.

(b) Find the marginal PMFs of X and Y.

(c) Find $P(Y = 2|X = 1)$.

(d) Are X and Y independent?

Solution:

(a)

$$
\begin{aligned}
P(X \leq 2, Y > 1) &= P(X = 1, Y = 2) + P(X = 2, Y = 2) \\
&= \frac{1}{12} + 0 = \frac{1}{12}.
\end{aligned}
$$

(b)

$$P_X(x) = \sum_{y \in R_Y} P(X = x, Y = y).$$

$$
P_X(x) = \begin{cases}
\frac{1}{3} + \frac{1}{12} = \frac{5}{12} & \text{for } x = 1 \\
\frac{1}{6} + 0 = \frac{1}{6} & \text{for } x = 2 \\
\frac{1}{12} + \frac{1}{3} = \frac{5}{12} & \text{for } x = 4
\end{cases}
$$

So:

$$
P_X(x) = \begin{cases}
\frac{5}{12} & x = 1 \\
\frac{1}{6} & x = 2 \\
\frac{5}{12} & x = 4
\end{cases}
$$

$$P_Y(y) = \sum_{x \in R_X} P(X = x, Y = y).$$

$$
P_Y(y) = \begin{cases}
\frac{1}{3} + \frac{1}{6} + \frac{1}{12} = \frac{7}{12} & \text{for } y = 1 \\
\frac{1}{12} + 0 + \frac{1}{3} = \frac{5}{12} & \text{for } y = 2
\end{cases}
$$

So:

$$P_Y(y) = \begin{cases} \frac{7}{12} & y = 1 \\ \\ \frac{5}{12} & y = 2 \end{cases}$$

(c)

$$P(Y = 2|X = 1) = \frac{P(Y = 2, X = 1)}{P(X = 1)} = \frac{\frac{1}{12}}{\frac{5}{12}} = \frac{1}{5}.$$

(d) Using the results of the previous part, we observe that:

$P(Y = 2|X = 1) = \frac{1}{5} \neq P(Y = 2) = \frac{5}{12}$.

So, we conclude that the two variables are not independent.

3. A box contains two coins: a regular coin and a biased coin with $P(H) = \frac{2}{3}$. I choose a coin at random and toss it once. I define the random variable X as a Bernoulli random variable associated with this coin toss, i.e., $X = 1$ if the result of the coin toss is heads and $X = 0$ otherwise. Then I take the remaining coin in the box and toss it once. I define the random variable Y as a Bernoulli random variable associated with the second coin toss. Find the joint PMF of X and Y. Are X and Y independent?

Solution:

We choose each coin with probability 0.5. We call the regular coin "coin1" and the biased coin "coin2."

Let X be a Bernoulli random variable associated with the first chosen coin toss. We can pick the first coin "coin1" or second coin "coin2" with equal probability 0.5. Thus, we can use the law of total probability:

$$\begin{aligned} P(X=1) &= P(\text{coin1})P(H|\text{coin } 1) + P(\text{coin2})P(H|\text{coin } 2) \\ &= \frac{1}{2}\times\frac{1}{2}+\frac{1}{2}\times\frac{2}{3} = \frac{7}{12}. \end{aligned}$$

$$\begin{aligned} P(X=0) &= P(\text{coin1})P(T|\text{coin } 1) + P(\text{coin2})P(T|\text{coin } 2) \\ &= \frac{1}{2}\times\frac{1}{2}+\frac{1}{2}\times\frac{1}{3} = \frac{5}{12}. \end{aligned}$$

Let Y be a Bernoulli random variable associated with the second chosen coin toss. We can pick the first coin "coin1" or second coin "coin2" with equal probability 0.5.

$$\begin{aligned} P(Y=1) &= P(\text{coin1})P(H|\text{coin } 1) + P(\text{coin2})P(H|\text{coin } 2) \\ &= \frac{1}{2}\times\frac{1}{2}+\frac{1}{2}\times\frac{2}{3} = \frac{7}{12}. \end{aligned}$$

$$\begin{aligned} P(Y=0) &= P(\text{coin1})P(T|\text{coin } 1) + P(\text{coin2})P(T|\text{coin } 2) \\ &= \frac{1}{2}\times\frac{1}{2}+\frac{1}{2}\times\frac{1}{3} = \frac{5}{12}. \end{aligned}$$

$$\begin{aligned}
P(X=0,Y=0) &= P(\text{first coin} = \text{coin1})P(T|\text{coin 1})P(T|\text{coin 2}) \\
&+ P(\text{first coin} = \text{coin2})P(T|\text{coin 1})P(T|\text{coin 2}) \\
&= P(T|\text{coin 1})P(T|\text{coin 2}) \\
&= \frac{1}{2} \times \frac{1}{3} = \frac{1}{6}.
\end{aligned}$$

$$\begin{aligned}
P(X=0,Y=1) &= P(\text{first coin} = \text{coin1})P(T|\text{coin 1})P(H|\text{coin 2}) \\
&+ P(\text{first coin} = \text{coin2})P(T|\text{coin 2})P(H|\text{coin 1}) \\
&= \frac{1}{2} \times \frac{1}{2} \times \frac{2}{3} + \frac{1}{2} \times \frac{1}{3} \times \frac{1}{2} = \frac{1}{4}.
\end{aligned}$$

$$\begin{aligned}
P(X=1,Y=0) &= P(\text{first coin} = \text{coin1})P(H|\text{coin 1})P(T|\text{coin 2}) \\
&+ P(\text{first coin} = \text{coin2})P(H|\text{coin 2})P(T|\text{coin 1}) \\
&= \frac{1}{2} \times \frac{1}{2} \times \frac{1}{3} + \frac{1}{2} \times \frac{2}{3} \times \frac{1}{2} = \frac{1}{4}.
\end{aligned}$$

$$\begin{aligned}
P(X=1,Y=1) &= P(\text{first coin} = \text{coin1})P(H|\text{coin 1})P(H|\text{coin 2}) \\
&+ P(\text{first coin} = \text{coin2})P(H|\text{coin 1})P(H|\text{coin 2}) \\
&= P(H|\text{coin 1})P(H|\text{coin 2}) \\
&= \frac{1}{2} \times \frac{2}{3} = \frac{1}{3}.
\end{aligned}$$

Table 5.2 summarizes the joint PMF of X and Y.

Table 5.2: Joint PMF of X and Y

	$Y=0$	$Y=1$
$X=0$	$\frac{1}{6}$	$\frac{1}{4}$
$X=1$	$\frac{1}{4}$	$\frac{1}{3}$

By comparing joint PMFs and marginal PMFs, we conclude that the two variables are not independent.

For example:

$$
\begin{aligned}
P(X=0) &= \frac{5}{12} \\
P(Y=1) &= \frac{7}{12} \\
P(X=0, Y=1) &= \frac{1}{4} \neq P(X=0) \times P(Y=1).
\end{aligned}
$$

5. Let X and Y be as defined in Problem 5. Also, suppose that we are given that $Y = 1$.

   (a) Find the conditional PMF of X given $Y = 1$. That is, find $P_{X|Y}(x|1)$.

   (b) Find $E[X|Y = 1]$.

   (c) Find $\text{Var}(X|Y = 1)$.

Solution:

(a)

$$
P_{X|Y}(x|1) = \frac{P(X=x, Y=1)}{P(Y=1)} = \frac{P(X=x, Y=1)}{\frac{7}{12}} = \frac{12}{7} P(X=x, Y=1).
$$

$$
P_{X|Y}(x|1) = \begin{cases} \frac{12}{7} \times \frac{1}{3} = \frac{4}{7} & x = 1 \\ \\ \frac{12}{7} \times \frac{1}{6} = \frac{2}{7} & x = 2 \\ \\ \frac{12}{7} \times \frac{1}{12} = \frac{1}{7} & x = 4 \end{cases}
$$

$$P_{X|Y}(x|1) = \begin{cases} \frac{4}{7} & x = 1 \\ \frac{2}{7} & x = 2 \\ \frac{1}{7} & x = 4 \end{cases}$$

(b)

$$E[X|Y=1] = \sum_x x P_{X|Y}(x|1) = 1 \times \frac{4}{7} + 2 \times \frac{2}{7} + 4 \times \frac{1}{7} = \frac{12}{7}.$$

(c)

$$E[X^2|Y=1] = \sum_x x^2 P_{X|Y}(x|1) = 1 \times \frac{4}{7} + 4 \times \frac{2}{7} + 16 \times \frac{1}{7} = \frac{28}{7}.$$

$$\begin{aligned} \text{Var}(X|Y=1) &= E(X^2|Y=1) - (E[X|Y=1])^2 \\ &= \frac{28}{7} - \left(\frac{12}{7}\right)^2 \\ &= \frac{52}{49} \end{aligned}$$

7. Let $X \sim$ *Geometric*(p). Find Var(X) as follows: find EX and EX^2 by conditioning on the result of the first "coin toss" and use Var(X)= $EX^2 - (EX)^2$.

Solution: The random experiment behind *Geometric*(p) is that we have a coin with $P(H) = p$. We toss the coin repeatedly until we observe the first heads. X is the total number of coin tosses. Now, there are two possible outcomes for the first coin toss: H or T. Thus, we can use the law of total expectation:

$$\begin{aligned} EX &= E[X|H]P(H) + E[X|T]P(T) \\ &= pE[X|H] + (1-p)E[X|T] \\ &= p \cdot 1 + (1-p)(EX+1). \end{aligned}$$

In this equation, $E[X|T] = 1 + EX$ because the tosses are independent, so if the first toss is tails, it is like starting over on the second toss. Solving for EX, we obtain

$$EX = \frac{1}{p}$$

Similarly, we can obtain EX^2.

$$\begin{aligned} EX^2 &= E[X^2|H]P(H) + E[X^2|T]P(T) \\ &= pE[X^2|H] + (1-p)E[X^2|T] \\ &= p \cdot 1 + (1-p)E(X+1)^2 \\ &= p + (1-p)[1 + 2EX + EX^2] \\ &= p + (1-p)\left[1 + \frac{2}{p} + EX^2\right] \end{aligned}$$

Solving for EX^2, we obtain

$$EX^2 = \frac{2-p}{p^2}.$$

Therefore,

$$\text{Var}(X) = EX^2 - (EX)^2 = \frac{1-p}{p^2}.$$

9. Consider the set of points in the set C:

$$C = \{(x,y) | x, y \in \mathbb{Z}, x^2 + |y| \leq 2\}.$$

Suppose that we pick a point (X, Y) from this set completely at random. Thus, each point has a probability of $\frac{1}{11}$ of being chosen.

(a) Find the joint and marginal PMFs of X and Y.

(b) Find the conditional PMF of X given $Y = 1$.

(c) Are X and Y independent?

(d) Find $E[XY^2]$.

Solution:

(a) Note that here

$$R_{XY} = C = \{(x,y)|x,y \in \mathbb{Z}, x^2 + |y| \leq 2\}.$$

Thus, the joint PMF is given by

$$P_{XY}(x,y) = \begin{cases} \frac{1}{11} & (x,y) \in C \\ 0 & \text{otherwise} \end{cases}$$

To find the marginal PMF of Y, $P_Y(j)$, we use

$$P_Y(y) = \sum_{x_i \in R_X} P_{XY}(x_i, y), \qquad \text{for any } y \in R_Y$$

Thus,

$$P_Y(-2) = P_{XY}(0,-2) = \frac{1}{11},$$
$$P_Y(-1) = P_{XY}(0,-1) + P_{XY}(-1,-1) + P_{XY}(1,-1) = \frac{3}{11},$$
$$P_Y(0) = P_{XY}(0,0) + P_{XY}(1,0) + P_{XY}(-1,0) = \frac{3}{11},$$
$$P_Y(1) = P_{XY}(0,1) + P_{XY}(-1,1) + P_{XY}(1,1) = \frac{3}{11},$$
$$P_Y(2) = P_{XY}(0,2) = \frac{1}{11}.$$

Similarly, we can find

$$P_X(i) = \begin{cases} \frac{3}{11} & \text{for } i = -1, 1 \\ \frac{5}{11} & \text{for } i = 0 \\ 0 & \text{otherwise} \end{cases}$$

(b) For $i = -1, 0, 1$, we can write

$$\begin{aligned} P_{X|Y}(i|1) &= \frac{P_{XY}(i,1)}{P_Y(1)} \\ &= \frac{\frac{1}{11}}{\frac{3}{11}} = \frac{1}{3}, \qquad \text{for } i = -1, 0, 1. \end{aligned}$$

Thus, we conclude

$$P_{X|Y}(i|1) = \begin{cases} \frac{1}{3} & \text{for } i = -1, 0, 1 \\ 0 & \text{otherwise} \end{cases}$$

By looking at the above conditional PMF, we conclude that given $Y = 1$, X is uniformly distributed over the set $\{-1, 0, 1\}$.

(c) X and Y are **not** independent. We can see this because the conditional PMF of X given $Y = 1$ (calculated above) is not the same as marginal PMF of X, $P_X(x)$.

(d) We have

$$\begin{aligned} E[XY^2] &= \sum_{i,j \in R_{XY}} ij^2 P_{XY}(i,j) \\ &= \frac{1}{11} \sum_{i,j \in R_{XY}} ij^2 \\ &= 0 \end{aligned}$$

11. The number of cars being repaired at a small repair shop has the following PMF:

$$P_N(n) = \begin{cases} \frac{1}{8} & \text{for } n = 0 \\ \frac{1}{8} & \text{for } n = 1 \\ \frac{1}{4} & \text{for } n = 2 \\ \frac{1}{2} & \text{for } n = 3 \\ 0 & \text{otherwise} \end{cases}$$

Each vehicle being repaired is a four-door car with probability $\frac{3}{4}$ and a two-door car with probability $\frac{1}{4}$ independently from other cars and independently from the total number of cars being repaired. Let X be the number of four-door cars and Y be the number of two-door cars currently being repaired.

(a) Find the marginal PMFs of X and Y.

(b) Find joint PMF of X and Y.

(c) Are X and Y independent?

Solution:

(a) Suppose that the number of cars being repaired is N. Then note that $R_X = R_Y = \{0, 1, 2, 3\}$ and $X + Y = N$. Also, given $N = n$, X is the sum of n independent $Bernoulli(\frac{3}{4})$ random variables. Thus, given $N = n$, X has a binomial distribution with parameters n and $\frac{3}{4}$, so

$$\begin{aligned} X|N = n &\sim Binomial(n, p = \frac{3}{4}); \\ Y|N = n &\sim Binomial(n, q = 1 - p = \frac{1}{4}). \end{aligned}$$

We have

$$\begin{aligned} P_X(k) &= \sum_{n=0}^{3} P(X = k|N = n)P_N(n) \quad \text{(law of total probability)} \\ &= \sum_{n=0}^{3} \binom{n}{k} p^k q^{n-k} P_N(n) \end{aligned}$$

$$P_X(k) = \begin{cases} \sum_{n=0}^{3} \binom{n}{0} \left(\frac{3}{4}\right)^0 \left(\frac{1}{4}\right)^n \cdot P_N(n) & \text{for } k = 0 \\ \sum_{n=0}^{3} \binom{n}{1} \left(\frac{3}{4}\right)^1 \left(\frac{1}{4}\right)^{n-1} \cdot P_N(n) & \text{for } k = 1 \\ \sum_{n=0}^{3} \binom{n}{2} \left(\frac{3}{4}\right)^2 \left(\frac{1}{4}\right)^{n-2} \cdot P_N(n) & \text{for } k = 2 \\ \sum_{n=0}^{3} \binom{n}{3} \left(\frac{3}{4}\right)^3 \left(\frac{1}{4}\right)^{n-3} \cdot P_N(n) & \text{for } k = 3 \\ 0 & \text{otherwise} \end{cases}$$

$$P_X(k) = \begin{cases} \frac{23}{128} & \text{for } k = 0 \\ \frac{33}{128} & \text{for } k = 1 \\ \frac{45}{128} & \text{for } k = 2 \\ \frac{27}{128} & \text{for } k = 3 \\ 0 & \text{otherwise} \end{cases}$$

Similarly, for the marginal PMF of Y, $p = \frac{1}{4}$ and $q = \frac{3}{4}$.

$$P_Y(k) = \begin{cases} \frac{73}{128} & \text{for } k = 0 \\ \frac{43}{128} & \text{for } k = 1 \\ \frac{11}{128} & \text{for } k = 2 \\ \frac{1}{128} & \text{for } k = 3 \\ 0 & \text{otherwise} \end{cases}$$

(b) To find the joint PMF of X and Y, we can also use the law of total probability:

$$P_{XY}(i,j) = \sum_{n=0}^{3} P(X = i, Y = j|N = n)P_N(n) \quad \text{(law of total probability)}.$$

But note that $P(X = i, Y = j|N = n) = 0$ if $N \neq i + j$, thus for $i, j \in \{0, 1, 2, 3\}$, we can write

$$\begin{aligned} P_{XY}(i,j) &= P(X = i, Y = j|N = i + j)P_N(i + j) \\ &= P(X = i|N = i + j)P_N(i + j) \\ &= \binom{i+j}{i}(\frac{3}{4})^i(\frac{1}{4})^j P_N(i + j) \end{aligned}$$

$$P_{XY}(i,j) = \begin{cases} \frac{1}{8}(\frac{3}{4})^i(\frac{1}{4})^j & \text{for } i + j = 0 \quad (i.e., i = j = 0) \\ \frac{1}{8}(\frac{3}{4})^i(\frac{1}{4})^j & \text{for } i + j = 1 \\ \frac{1}{4}\binom{2}{i}(\frac{3}{4})^i(\frac{1}{4})^j & \text{for } i + j = 2 \\ \frac{1}{2}\binom{3}{i}(\frac{3}{4})^i(\frac{1}{4})^j & \text{for } i + j = 3 \\ 0 & \text{otherwise} \end{cases}$$

(c) X and Y are **not** independent since, as we saw above:

$$P_{XY}(i,j) \neq P_X(i)P_Y(j).$$

13. Consider two random variables X and Y with their joint PMF given in Table 5.5.

Table 5.3: Joint PMF of X and Y in Problem 13.

	$Y = 0$	$Y = 1$	$Y = 2$
$X = 0$	$\frac{1}{6}$	$\frac{1}{6}$	$\frac{1}{8}$
$X = 1$	$\frac{1}{8}$	$\frac{1}{6}$	$\frac{1}{4}$

Define the random variable Z as $Z = E[X|Y]$.

(a) Find the marginal PMFs of X and Y.

(b) Find the conditional PMF of X given $Y = 0$ and $Y = 1$, i.e., find $P_{X|Y}(x|0)$ and $P_{X|Y}(x|1)$.

(c) Find the PMF of Z.

(d) Find EZ and check that $EZ = EX$.

(e) Find $\text{Var}(Z)$.

Solution:

(a) Using the table, we find out

$$\begin{aligned} P_X(0) &= \frac{1}{6} + \frac{1}{6} + \frac{1}{8} = \frac{11}{24}, \\ P_X(1) &= \frac{1}{8} + \frac{1}{6} + \frac{1}{4} = \frac{13}{24}, \\ P_Y(0) &= \frac{1}{6} + \frac{1}{8} = \frac{7}{24}, \\ P_Y(1) &= \frac{1}{6} + \frac{1}{6} = \frac{1}{3}, \\ P_Y(2) &= \frac{1}{8} + \frac{1}{4} = \frac{3}{8}. \end{aligned}$$

Note that X and Y are not independent.

(b) We have

$$\begin{aligned} P_{X|Y}(0|0) &= \frac{P_{XY}(0,0)}{P_Y(0)} \\ &= \frac{\frac{1}{6}}{\frac{7}{24}} = \frac{4}{7}. \end{aligned}$$

Thus,

$$P_{X|Y}(1|0) = 1 - \frac{4}{7} = \frac{3}{7}.$$

We conclude

$$X|Y = 0 \;\sim\; Bernoulli\left(\frac{3}{7}\right).$$

Similarly, we find

$$\begin{aligned} P_{X|Y}(0|1) &= \frac{1}{2}, \\ P_{X|Y}(1|1) &= \frac{1}{2}. \end{aligned}$$

(c) We note that the random variable Y can take three values: 0, 1, and 2. Thus, the random variable $Z = E[X|Y]$ can take three values as it is a function of Y. Specifically,

$$Z = E[X|Y] = \begin{cases} E[X|Y=0] & \text{if } Y=0 \\ \\ E[X|Y=1] & \text{if } Y=1 \\ \\ E[X|Y=2] & \text{if } Y=2 \end{cases}$$

Now, using the previous part, we have

$$E[X|Y=0] = \frac{3}{7}, \quad E[X|Y=1] = \frac{1}{2}, \quad E[X|Y=2] = \frac{2}{3}$$

and since $P(Y=0) = \frac{7}{24}$, $P(Y=1) = \frac{1}{3}$, and $P(Y=2) = \frac{3}{8}$ we conclude that

$$Z = E[X|Y] = \begin{cases} \frac{3}{7} & \text{with probability } \frac{7}{24} \\ \\ \frac{1}{2} & \text{with probability } \frac{1}{3} \\ \\ \frac{2}{3} & \text{with probability } \frac{3}{8} \end{cases}$$

So we can write

$$P_Z(z) = \begin{cases} \frac{7}{24} & \text{if } z=\frac{3}{7} \\ \\ \frac{1}{3} & \text{if } z=\frac{1}{2} \\ \\ \frac{3}{8} & \text{if } z=\frac{2}{3} \\ \\ 0 & \text{otherwise} \end{cases}$$

(d) Now that we have found the PMF of Z, we can find its mean and variance. Specifically,

$$E[Z] = \frac{3}{7} \cdot \frac{7}{24} + \frac{1}{2} \cdot \frac{1}{3} + \frac{2}{3} \cdot \frac{3}{8} = \frac{13}{24}.$$

We also note that $EX = \frac{13}{24}$. Thus, here we have

$$E[X] = E[Z] = E[E[X|Y]].$$

(e) To find $\text{Var}(Z)$, we write

$$\begin{aligned} \text{Var}(Z) &= E[Z^2] - (EZ)^2 \\ &= E[Z^2] - (\frac{13}{24})^2, \end{aligned}$$

where

$$E[Z^2] = (\frac{3}{7})^2 \cdot \frac{7}{24} + (\frac{1}{2})^2 \cdot \frac{1}{3} + (\frac{2}{3})^2 \cdot \frac{3}{8} = \frac{17}{56}.$$

Thus,

$$\begin{aligned}\text{Var}(Z) &= \frac{17}{56} - (\frac{13}{24})^2 \\ &= \frac{41}{4032}.\end{aligned}$$

15. Let N be the number of phone calls made by the customers of a phone company in a given hour. Suppose that $N \sim Poisson(\beta)$, where $\beta > 0$ is known. Let X_i be the length of the i'th phone call, for $i = 1, 2, ..., N$. We assume X_i's are independent of each other and also independent of N. We further assume

$$X_i \sim Exponential(\lambda),$$

where $\lambda > 0$ is known. Let Y be the sum of the lengths of the phone calls, i.e.,

$$Y = \sum_{i=1}^{N} X_i.$$

Find EY and $\text{Var}(Y)$.

Solution: To find EY, we cannot directly use the linearity of expectation because N is random but, conditioned on $N = n$, we can use linearity and find $E[Y|N = n]$; so, we use the law of iterated expectations:

$$\begin{aligned}
EY &= E[E[Y|N]] && \text{(law of iterated expectations)}\\
&= E\left[E\left[\sum_{i=1}^{N} X_i|N\right]\right]\\
&= E\left[\sum_{i=1}^{N} E[X_i|N]\right] && \text{(linearity of expectation)}\\
&= E\left[\sum_{i=1}^{N} E[X_i]\right] && \text{(X_i's and N are indpendent)}\\
&= E[NE[X]] && \text{(since $EX_i = EX$s)}\\
&= E[X]E[N] && \text{(since EX is not random).}\\
EY &= E[X]E[N]\\
EY &= \frac{1}{\lambda}\cdot\beta\\
EY &= \frac{\beta}{\lambda}
\end{aligned}$$

To find $\text{Var}(Y)$, we use the law of total variance:

$$\begin{aligned}
\text{Var}(Y) &= E(\text{Var}(Y|N)) + \text{Var}(E[Y|N])\\
&= E(\text{Var}(Y|N)) + \text{Var}(NEX) && \text{(as above)}\\
&= E(\text{Var}(Y|N)) + (EX)^2\text{Var}(N).
\end{aligned}$$

To find $E(\text{Var}(Y|N))$ note that, given $N = n$, Y is the sum of n independent random variables. As we discussed before, for n independent random variables, the variance of the sum is equal to sum of the variances. We can write

$$\begin{aligned}
\text{Var}(Y|N) &= \sum_{i=1}^{N} \text{Var}(X_i|N)\\
&= \sum_{i=1}^{N} \text{Var}(X_i) && \text{(since X_i's are independent of N)}\\
&= N\text{Var}(X).
\end{aligned}$$

Thus, we have

$$E(\mathrm{Var}(Y|N)) = EN\mathrm{Var}(X).$$

We obtain

$$\begin{aligned}\mathrm{Var}(Y) &= ENVar(X) + (EX)^2\mathrm{Var}(N).\\ \mathrm{Var}(Y) &= \beta(\frac{1}{\lambda})^2 + (\frac{1}{\lambda})^2\beta.\\ &= \left(\frac{2\beta}{\lambda^2}\right)\end{aligned}$$

17. Let X and Y be two jointly continuous random variables with joint PDF

$$f_{XY}(x,y) = \begin{cases} e^{-xy} & 1 \le x \le e,\ y > 0 \\ 0 & \text{otherwise} \end{cases}$$

(a) Find the marginal PDFs, $f_X(x)$ and $f_Y(y)$.

(b) Write an integral to compute $P(0 \le Y \le 1, 1 \le X \le \sqrt{e})$.

Solution:

(a) We have:

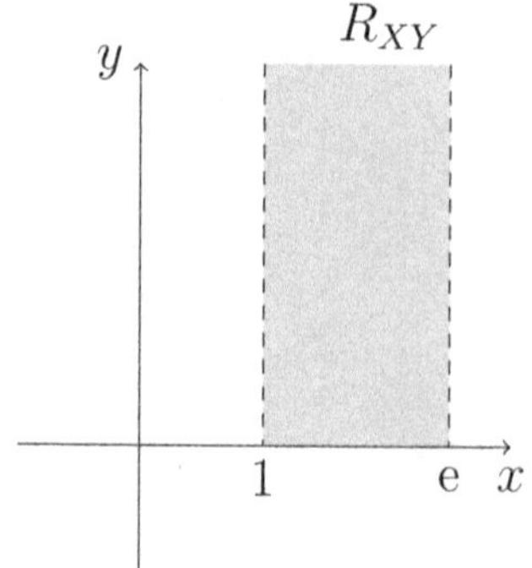

for $1 < x < e$:

$$f_X(x) = \int_0^\infty e^{-xy} dy$$
$$= -\frac{1}{x} e^{-xy} \Big|_0^\infty$$
$$= \frac{1}{x}$$

$$f_X(x) = \begin{cases} \frac{1}{x} & 1 \leq x \leq e \\ 0 & \text{otherwise} \end{cases}$$

for $0 < y$

$$f_Y(y) = \int_1^e e^{-xy} dx$$
$$= \frac{1}{y}(e^{-y} - e^{-ey})$$

Thus,

$$f_Y(y) = \begin{cases} \frac{1}{y}(e^{-y} - e^{-ey}) & y > 0 \\ 0 & \text{otherwise} \end{cases}$$

(b)

$$P(0 \leq Y \leq 1, 1 \leq X \leq \sqrt{e}) = \int_{x=1}^{\sqrt{e}} \int_{y=0}^{1} e^{-xy} dy dx$$
$$= \frac{1}{2} - \int_1^{\sqrt{e}} \frac{1}{x} e^{-x} dx$$

19. Let X and Y be two jointly continuous random variables with joint CDF

$$F_{XY}(x, y) = \begin{cases} 1 - e^{-x} - e^{-2y} + e^{-(x+2y)} & x, y > 0 \\ 0 & \text{otherwise} \end{cases}$$

(a) Find the joint PDF, $f_{XY}(x, y)$.

(b) Find $P(X < 2Y)$.

(c) Are X and Y independent?

Solution: Note that we can write $F_{XY}(x, y)$ as

$$\begin{aligned} F_{XY}(x, y) &= \left(1 - e^{-x}\right) u(x)(1 - e^{-2y})u(y) \\ &= (\text{a function of} \quad x) \cdot (\text{a function of} \quad y) \\ &= F_X(x) \cdot F_Y(y) \end{aligned}$$

i.e. X and Y are independent.

(a)

$$F_X(x) = (1 - e^{(-x)})u(x)$$

Thus $X \sim \mathit{Exponential}(1)$. So, we have $f_X(x) = e^{-x}u(x)$. Similarly, $f_Y(y) = 2e^{-2y}u(y)$ which results in:

$$f_{XY}(x, y) = 2e^{(-x+2y)}u(x)u(y)$$

(b)

$$\begin{aligned} P(X < 2Y) &= \int_{y=0}^{\infty} \int_{x=0}^{2y} 2e^{-(x+2y)} dx dy \\ &= \int_{y=0}^{\infty} \left(2e^{-2y} - 2e^{-4y}\right) dy \\ &= \frac{1}{2} \end{aligned}$$

(c) Yes, as we saw above.

21. Let X and Y be two jointly continuous random variables with joint PDF

$$f_{XY}(x,y) = \begin{cases} x^2 + \frac{1}{3}y & -1 \le x \le 1, 0 \le y \le 1 \\ 0 & \text{otherwise} \end{cases}$$

(a) Find he conditional PDF of X given $Y = y$, for $0 \le y \le 1$.

(b) Find $P(X > 0|Y = y)$, for $0 \le y \le 1$. Does this value depend on y?

(c) Are X and Y independent?

Solution:

(a) Let us first find $f_Y(y)$:

$$f_Y(y) = \int_{-1}^{+1} (x^2 + \frac{1}{3}y)dx = \left[\frac{1}{3}x^3 + \frac{1}{3}yx\right]_{-1}^{+1}$$
$$= \frac{2}{3}y + \frac{2}{3} \quad \text{for} \quad 0 \le y \le 1$$

Thus, for $0 \le y \le 1$, we obtain:

$$f_{X|Y}(x|y) = \frac{f_{XY}(x,y)}{f_Y(y)} = \frac{x^2 + \frac{1}{3}y}{\frac{2}{3}y + \frac{2}{3}} = \frac{3x^2 + y}{2y + 2} \quad \text{for} \quad -1 \le x \le 1$$

For $0 \le y \le 1$:

$$f_{X|Y}(x|y) = \begin{cases} \frac{3x^2+y}{2y+2} & -1 \le x \le 1 \\ 0 & \text{else} \end{cases}$$

(b)

$$P(X > 0|Y = y) = \int_0^1 f_{X|Y}(x|y)dx = \int_0^1 \frac{3x^2 + y}{2y + 2}dx$$
$$= \frac{1}{2y + 2}\int_0^1 (3x^2 + y)dx$$
$$= \frac{1}{2y + 2}\left[(x^3 + yx)\right]_0^1 = \frac{y + 1}{2(y + 1)} = \frac{1}{2}$$

Thus it does not depend on y.

(c) X and Y are not independent. Since $f_{X|Y}(x|y)$ depends on y.

23. Consider the set

$$E = \{(x, y) \big| |x| + |y| \leq 1\}.$$

Suppose that we choose a point (X, Y) uniformly at random in E. That is, the joint PDF of X and Y is given by

$$f_{XY}(x, y) = \begin{cases} c & (x, y) \in E \\ \\ 0 & \text{otherwise} \end{cases}$$

(a) Find the constant c.

(b) Find the marginal PDFs $f_X(x)$ and $f_Y(y)$.

(c) Find the conditional PDF of X given $Y = y$, where $-1 \leq y \leq 1$.

(d) Are X and Y independent?

Solution:

(a) We have:

$$1 = \int_E \int c\,dx\,dy = c(\text{area of E}) = c\sqrt{2} \cdot \sqrt{2} = 2c$$
$$\rightarrow c = \frac{1}{2}$$

(b)

For $0 \leq x \leq 1$, we have:

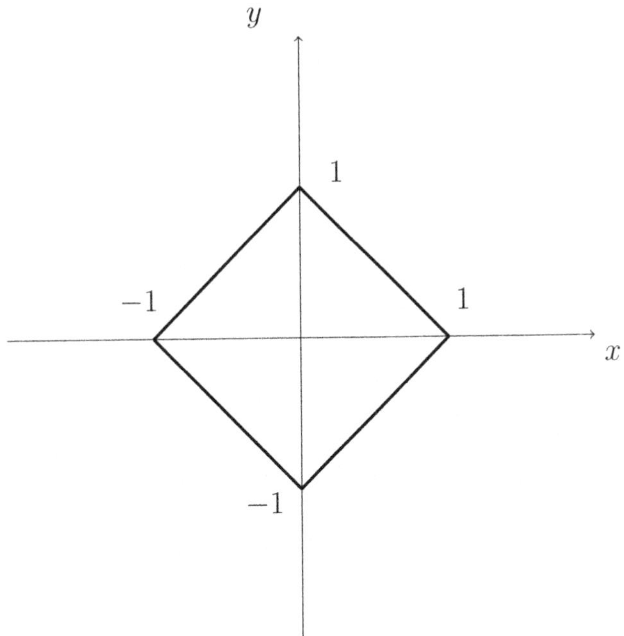
y
1
−1
1
x
−1

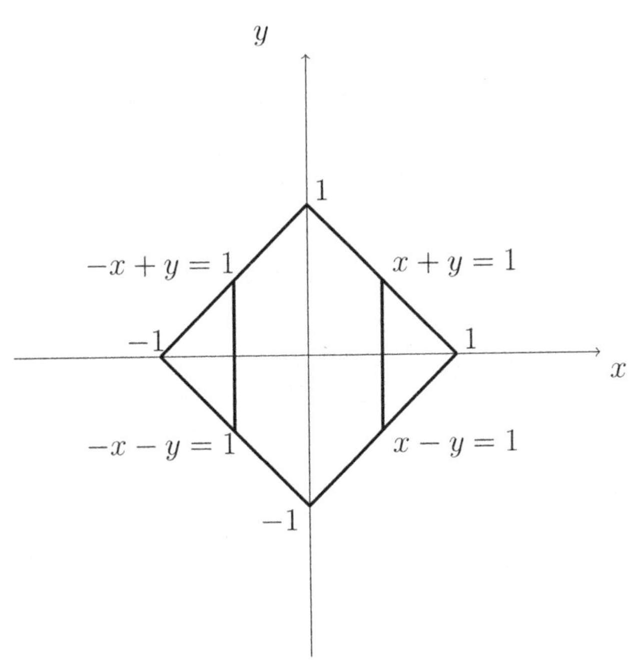
y
1
−x + y = 1
x + y = 1
−1
1
x
−x − y = 1
x − y = 1
−1

$$f_X(x) = \int_{x-1}^{1-x} \frac{1}{2} dy = 1 - x$$

For $-1 \leq x \leq 0$, we have:

$$f_X(x) = \int_{-x-1}^{1+x} \frac{1}{2} dy = 1 + x$$

$$f_X(x) = \begin{cases} 1 - |x| & -1 \leq x \leq 1 \\ \\ 0 & \text{else} \end{cases}$$

Similarly, we find:

$$f_Y(y) = \begin{cases} 1 - |y| & -1 \leq y \leq 1 \\ \\ 0 & \text{else} \end{cases}$$

(c)

$$\begin{aligned} f_{X|Y}(x|y) &= \frac{f_{XY}(xy)}{f_Y(y)} = \frac{\frac{1}{2}}{(1 - |y|)} \\ &= \frac{1}{2(1 - |y|)} \quad \text{for } |x| \leq 1 - |y| \end{aligned}$$

Thus:

$$f_{X|Y}(x|y) = \begin{cases} \frac{1}{2(1-|y|)} & \text{for } -1 + |y| \leq x \leq 1 - |y| \\ \\ 0 & \text{else} \end{cases}$$

So, we conclude that given $Y = y$, X is uniformly distributed on $[-1 + |y|, 1 - |y|]$, i.e.:

$X|Y = y \sim Uniform(-1 + |y|, 1 - |y|)$
(d) No, because $f_{XY}(x, y) \neq f_X(x) \cdot f_Y(y)$

25. Suppose $X \sim Exponential(1)$ and given $X = x$, Y is a uniform random variable in $[0., x]$, i.e.,

$$Y|X = x \quad \sim \quad Uniform(0, x),$$

or equivalently

$$Y|X \quad \sim \quad Uniform(0, X).$$

(a) Find EY.

(b) Find $\text{Var}(Y)$.

Solution:

Remember that if $Y \sim Uniform(a, b)$, then $EY = \frac{a+b}{2}$ and $\text{Var}(Y) = \frac{(b-a)^2}{12}$

(a) Using the law of total expectation:

$$\begin{aligned}
E[Y] &= \int_0^\infty E[Y|X = x] f_X(x) dx \\
&= \int_0^\infty E[Y|X = x] e^{-x} dx \quad \text{Since } Y|X \sim Uniform(0, X) \\
&= \int_0^\infty \frac{x}{2} e^{-x} dx = \frac{1}{2} [\int_0^\infty x e^{-x} dx] \\
&= \frac{1}{2} \cdot 1 = \frac{1}{2}
\end{aligned}$$

(b)

$$\begin{aligned}
EY^2 &= \int_0^\infty E[Y^2|X = x] f_X(x) dx \\
&= \int_0^\infty E[Y^2|X = x] e^{-x} dx \quad \text{Law of total expectation}
\end{aligned}$$

$Y|X \sim Uniform(0, X)$

$$\begin{aligned}
E[Y^2|X = x] &= \text{Var}(Y|X = x) + (E[Y|X = x])^2 \\
&= \frac{x^2}{12} + \frac{x^2}{4} = \frac{x^2}{3} \\
EY^2 &= \int_0^\infty \frac{x^2}{3} e^{-x} dx = \frac{1}{3} \int_0^\infty x^2 e^{-x} dx \\
&= \frac{1}{3} EW^2 = \frac{1}{3} [\text{Var}(W) + (EW)^2] \\
&= \frac{1}{3}(1 + 1) = \frac{2}{3} \quad \text{where } W \sim Exponential(1)
\end{aligned}$$

Therefore:

$$EY^2 = \frac{2}{3} \quad \text{Var}(Y) = \frac{2}{3} - \frac{1}{4} = \frac{5}{12}$$

27. Let X and Y be two independent $Uniform(0, 1)$ random variables and $Z = \frac{X}{Y}$. Find both the CDF and PDF of Z.

Solution:

First note that since $R_X = R_Y = [0, 1]$, we conclude $R_Z = [0, \infty)$. We first find the CDF of Z.

$$\begin{aligned}
F_Z(z) &= P(Z \leq z) = P(\frac{X}{Y} \leq z) \\
&= P(X \leq zY) \quad \text{(Since } Y \geq 0\text{)} \\
&= \int_0^1 P(X \leq zY|Y = y) f_Y(y) dy \quad \text{(Law of total prob)} \\
&= \int_0^1 P(X \leq zy) dy \quad \text{(Since } X \text{ and } Y \text{ are indep)}
\end{aligned}$$

Note:

$$P(X \leq zy) = \begin{cases} 1 & \text{if } y > \frac{1}{z} \\ zy & \text{if } y \leq \frac{1}{z} \end{cases}$$

Consider two cases:

(a) If $0 \leq z \leq 1$, then $P(X \leq zy) = zy$ for all $0 \leq y \leq 1$

Thus:

$$F_Z(z) = \int_0^1 (zy)dy = \frac{1}{2}zy^2|_0^1 = \frac{1}{2}z$$

(b) If $z > 1$, then

$$\begin{aligned} F_Z(z) &= \int_0^{\frac{1}{z}} zydy + \int_{\frac{1}{z}}^1 1dy \\ &= [\frac{1}{2}zy^2]_0^{\frac{1}{z}} + [y]_{\frac{1}{z}}^1 \\ &= \frac{1}{2z} + 1 - \frac{1}{z} = 1 - \frac{1}{2z} \end{aligned}$$

$$F_Z(z) = \begin{cases} \frac{1}{2}z & 0 \leq z \leq 1 \\ 1 - \frac{1}{2z} & z \geq 1 \\ 0 & z < 0 \end{cases}$$

Note that $F_Z(z)$ is a continuous function.

$$f_Z(z) = \frac{d}{dz}F_Z(z) = \begin{cases} \frac{1}{2} & 0 \leq z \leq 1 \\ \frac{1}{2z^2} & z \geq 1 \\ 0 & \text{else} \end{cases}$$

29. Let X and Y be two independent standard normal random variables. Consider the point (X, Y) in the $x - y$ plane. Let (R, Θ) be the corresponding polar coordinates as shown in Figure 5.11. The inverse transformation is given by

$$\begin{cases} X = R\cos\Theta \\ Y = R\sin\Theta \end{cases}$$

where, $R \geq 0$ and $-\pi < \Theta \leq \pi$. Find the joint PDF of R and Θ. Show that R and Θ are independent.

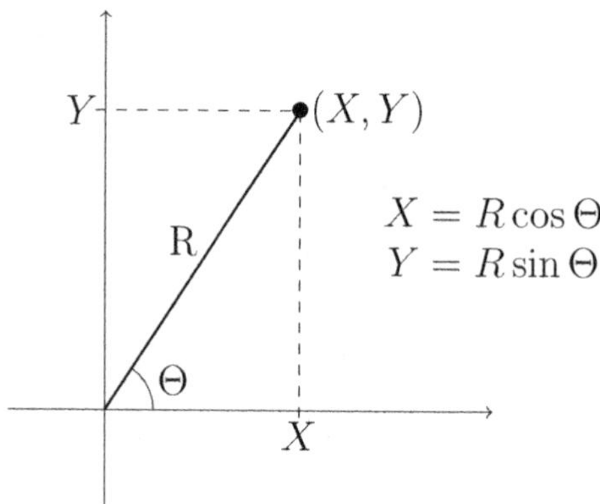

Figure 5.1: Polar coordinates

Solution: Here (X, Y) are jointly continuous with

$$f_{XY}(x, y) = \frac{1}{2\pi}e^{-\frac{x^2+y^2}{2}}.$$

Also, (X, Y) is related to (R, Θ) by a one-to-one relationship. We can use the method of transformations. The function $h(r, \theta)$ is given by

$$\begin{cases} x = h_1(r, \theta) = r\cos\theta \\ y = h_2(r, \theta) = r\sin\theta \end{cases}$$

Thus, we have

$$\begin{aligned} f_{R\Theta}(r, \theta) &= f_{XY}(h_1(r, \theta), h_2(r, \theta))|J| \\ &= f_{XY}(r\cos\theta, r\sin\theta)|J|. \end{aligned}$$

where

$$J = \det \begin{bmatrix} \frac{\partial h_1}{\partial r} & \frac{\partial h_1}{\partial \theta} \\ \\ \frac{\partial h_2}{\partial r} & \frac{\partial h_2}{\partial \theta} \end{bmatrix} = \det \begin{bmatrix} \cos\theta & -r\sin\theta \\ \\ \sin\theta & r\cos\theta \end{bmatrix} = r\cos^2\theta + r\sin^2\theta = r.$$

We conclude that

$$\begin{aligned} f_{R\Theta}(r,\theta) &= f_{XY}(r\cos\theta, r\sin\theta)|J| \\ &= \begin{cases} \frac{r}{2\pi}e^{-\frac{r^2}{2}} & r \in [0,\infty), \theta \in (-\pi,\pi] \\ \\ 0 & \text{otherwise} \end{cases} \end{aligned}$$

Note that, from above, we can write

$$f_{R\Theta}(r,\theta) = f_R(r)f_\Theta(\theta),$$

where

$$f_R(r) = \begin{cases} re^{-\frac{r^2}{2}} & r \in [0,\infty) \\ \\ 0 & \text{otherwise} \end{cases}$$

$$f_\Theta(\theta) = \begin{cases} \frac{1}{2\pi} & \theta \in (-\pi,\pi] \\ \\ 0 & \text{otherwise} \end{cases}$$

Thus, we conclude that R and Θ are independent.

31. Consider two random variables X and Y with joint PMF given in Table 5.6. Find $\text{Cov}(X,Y)$ and $\rho(X,Y)$.

Solution:

First, we find the PMFs of X and Y:
$R_X = \{0,1\}$ $\quad P_X(0) = \frac{1}{6}+\frac{1}{4}+\frac{1}{8} = \frac{4+6+3}{24} = \frac{13}{24}$ $\quad P_X(1) = \frac{1}{8}+\frac{1}{6}+\frac{1}{6} = \frac{11}{24}$
$R_Y = \{0,1,2\}$ $\quad P_Y(0) = \frac{1}{6}+\frac{1}{8} = \frac{7}{24}$
$P_Y(1) = \frac{1}{4}+\frac{1}{6} = \frac{5}{12}$ $\quad P_Y(2) = \frac{1}{8}+\frac{1}{6} = \frac{7}{24}$

Table 5.4: Joint PMF of X and Y in Problem 31.

	$Y=0$	$Y=1$	$Y=2$
$X=0$	$\frac{1}{6}$	$\frac{1}{4}$	$\frac{1}{8}$
$X=1$	$\frac{1}{8}$	$\frac{1}{6}$	$\frac{1}{6}$

$$\begin{aligned}
EX &= 0 \cdot \frac{13}{24} + 1 \cdot \frac{11}{24} = \frac{11}{24} \\
EY &= 0 \cdot \frac{7}{24} + 1 \cdot \frac{5}{12} + 2 \cdot \frac{7}{24} = 1 \\
EXY &= \sum ijP_{XY}(i,j) = 0 + 1 \cdot 0 \cdot \frac{1}{8} + 1 \cdot 1 \cdot \frac{1}{6} + 1 \cdot 2 \cdot \frac{1}{6} = \frac{1}{6} + \frac{1}{3} = \frac{1}{2}
\end{aligned}$$

Therefore:

$$\text{Cov}(X,Y) = EXY - EX \cdot EY = \frac{1}{2} - \frac{11}{24} \cdot 1 = \frac{1}{24}$$

$$\begin{aligned}
\text{Var}(X) &= EX^2 - (EX)^2 \\
EX^2 &= \sum i^2 P_{XY}(i,j) = \frac{11}{24} \\
\text{Var}(X) &= \frac{11}{24} \cdot \frac{13}{24} \\
\rightarrow \sigma_X &= \frac{\sqrt{11 \times 13}}{24} \approx 0.498
\end{aligned}$$

$$\begin{aligned}
EY^2 &= 0 \cdot \frac{7}{24} + 1 \cdot \frac{5}{12} + 4 \cdot \frac{7}{24} = \frac{19}{12} \\
\text{Var}(Y) &= \frac{19}{12} - 1 = \frac{7}{12} \\
\rightarrow \sigma_Y &= \sqrt{\frac{7}{12}} \approx 0.76
\end{aligned}$$

$$\begin{aligned}
\rightarrow \rho(X,Y) &= \frac{\text{Cov}(X,Y)}{\sigma_X \sigma_Y} \\
&= \frac{\frac{1}{24}}{\frac{\sqrt{11\times 13}}{24} \cdot \sqrt{\frac{7}{12}}} \approx 0.11
\end{aligned}$$

33. Let X and Y be two random variables. Suppose that $\sigma_X^2 = 4$, and $\sigma_Y^2 = 9$. If we know that the two random variables $Z = 2X - Y$ and $W = X + Y$ are independent, find $\text{Cov}(X,Y)$ and $\rho(X,Y)$.

Solution:

Z and W are independent, thus $\text{Cov}(Z,W) = 0$. Therefore:

$$\begin{aligned}
0 &= \text{Cov}(Z,W) = \text{Cov}(2X - Y, X + Y) \\
&= 2 \cdot \text{Var}(X) + 2 \cdot \text{Cov}(X,Y) - \text{Cov}(Y,X) - \text{Var}(Y) \\
&= 2 \times 4 + \text{Cov}(X,Y) - 9
\end{aligned}$$

Therefore:

$$\begin{aligned}
\text{Cov}(X,Y) &= 1 \\
\rho(X,Y) &= \frac{\text{Cov}(X,Y)}{\sigma_X \sigma_Y} \\
&= \frac{1}{2 \times 3} = \frac{1}{6}
\end{aligned}$$

35. Let X and Y be two independent $N(0,1)$ random variables and

$$\begin{aligned}
Z &= 7 + X + Y \\
W &= 1 + Y.
\end{aligned}$$

Find $\rho(Z, W)$.

Solution:

$$\begin{aligned}\text{Cov}(Z, W) &= \text{Cov}(7 + X + Y, 1 + Y) \\ &= \text{Cov}(X + Y, Y) \\ &= \text{Cov}(X, Y) + \text{Var}(Y).\end{aligned}$$

Since X and Y are independent, $\text{Cov}(X, Y) = 0$, so

$$\text{Cov}(Z, W) = \text{Var}(Y) = 1$$

$$\begin{aligned}\text{Var}(Z) &= \text{Var}(X + Y) \quad \text{Since } X \text{and } Y \text{ are independent} \\ &= \text{Var}(X) + \text{Var}(Y) = 2 \\ \text{Var}(W) &= \text{Var}(Y) = 1\end{aligned}$$

Therefore:

$$\begin{aligned}\rho(X, Y) &= \frac{\text{Cov}(Z, W)}{\sigma_Z \sigma_W} \\ &= \frac{1}{\sqrt{1 \times 2}} = \frac{1}{\sqrt{2}}\end{aligned}$$

37. Let X and Y be jointly normal random variables with parameters $\mu_X = 1$, $\sigma_X^2 = 4$, $\mu_Y = 1$, $\sigma_Y^2 = 1$, and $\rho = 0$.

   (a) Find $P(X + 2Y > 4)$.

   (b) Find $E[X^2 Y^2]$.

Solution:
$X \sim N(1,4)$; $Y \sim N(1,1)$:

$\rho(X,Y) = 0$ and X , Y are jointly normal. Therefore X and $Y$are independent.

(a) $W = X + 2Y$ Therefore:

$W \sim N(3, 4+4) = N(3,8)$

$$P(W > 4) = 1 - \Phi(\frac{4-3}{\sqrt{8}}) = 1 - \Phi(\frac{1}{\sqrt{8}})$$

(b)

$$\begin{aligned} E[X^2Y^2] &= EX^2 \cdot EY^2 \quad \text{Since } X \text{and } Y \text{ are independent.} \\ &= (4+1) \cdot (1+1) = 10 \end{aligned}$$

Chapter 6

Multiple Random Variables

1. Let X, Y, and Z be three jointly continuous random variables with joint PDF

$$f_{XYZ}(x,y,z) = \begin{cases} x+y & 0 \leq x,y,z \leq 1 \\ 0 & \text{otherwise} \end{cases}$$

   (a) Find the joint PDF of X and Y.

   (b) Find the marginal PDF of X.

   (c) Find the conditional PDF of $f_{XY|Z}(x,y|z)$ using

$$f_{XY|Z}(x,y|z) = \frac{f_{XYZ}(x,y,z)}{f_Z(z)}.$$

   (d) Are X and Y independent of Z?

Solution:

$$f_{XYZ}(x,y,z) = \begin{cases} x+y & 0 \leq x,y,z \leq 1 \\ 0 & \text{otherwise} \end{cases}$$

(a)

$$\begin{aligned} f_{XY}(x,y) &= \int_{-\infty}^{\infty} f_{XYZ}(x,y,z)dz \\ &= \int_0^1 (x+y)dz \\ &= x+y \end{aligned}$$

Thus,

$$f_{XY}(x,y) = \begin{cases} x+y & 0 \leq x,y \leq 1 \\ 0 & \text{otherwise} \end{cases}$$

(b)

$$\begin{aligned} f_X(x) &= \int_0^1 f_{XY}(x,y)dy \\ &= \int_0^1 (x+y)dy \\ &= \left[xy + \frac{1}{2}y^2\right]_0^1 \\ &= x + \frac{1}{2} \end{aligned}$$

$$f_X(x) = \begin{cases} x+\frac{1}{2} & 0 \leq x \leq 1 \\ 0 & \text{otherwise} \end{cases}$$

(c)

$$
\begin{aligned}
f_{XY|Z}(x, y, z) &= \frac{f_{XYZ}(x, y, z)}{f_Z(z)} \\
&= \frac{x+y}{f_Z(z)} \quad \text{for} \quad 0 \leq x, y, z \leq 1 \\
f_Z(z) &= \int_0^1 \int_0^1 (x+y) dy dx \\
&= \int_0^1 \left[xy + \frac{1}{2} y^2 \right]_0^1 dx \\
&= \int_0^1 (x + \frac{1}{2}) dx \\
&= \frac{1}{2} x^2 + \frac{1}{2} \Big|_0^1 \\
&= 1
\end{aligned}
$$

Thus,

$$
f_Z(z) = 1 \quad \text{for} \quad 0 < z < 1
$$

Thus,

$$
\begin{aligned}
f_{XY|Z}(x, y|z) &= x + y \\
&= f_{XY}(x, y) \quad \text{for} \quad 0 \leq x, y \leq 1
\end{aligned}
$$

(d) Yes, since $f_{XY|Z}(x, y|z) = f_{XY}(x, y)$. Also, note that $f_{XYZ}(x, y, z)$ can be written as a function of x, y times a function of z:

$$
f_{XYZ}(x, y, z) = h(x, y) g(z)
$$

where

$$
h(x, y) = \begin{cases} x + y & 0 \leq x, y \leq 1 \\ 0 & \text{otherwise} \end{cases}
$$

$$
g(z) = \begin{cases} 1 & 0 \leq z \leq 1 \\ 0 & \text{otherwise} \end{cases}
$$

3. Let X, Y, and Z be three independent $N(1,1)$ random variables. Find $E[XY|Y+Z=1]$.

   Solution:

$$\begin{aligned} E[XY|Y+Z=1] &= E[X]E[Y|Y+Z=1] \\ &= E[Y|Y+Z=1] \end{aligned}$$

   But note:

$$\begin{aligned} E[Y|Y+Z=1] &= E[Z|Y+Z=1] \quad \text{(by symmetry)} \\ E[Y|Y+Z=1]+E[Z|Y+Z=1] &= E[Y+Z|Y+Z=1] \\ &= 1 \end{aligned}$$

   Therefore,

$$\begin{aligned} E[Y|Y+Z=1] &= \frac{1}{2} \\ E[XY|Y+Z=1] &= \frac{1}{2} \end{aligned}$$

5. In this problem, our goal is to find the variance of the hypergeometric distribution. Let's remember the random experiment behind the hypergeometric distribution. Say you have a bag that contains b blue marbles and r red marbles. You choose $k \leq b+r$ marbles at random (without replacement) and let X be the number of blue marbles in your sample. Then $X \sim Hypergeometric(b,r,k)$. Now let us define the indicator random variables X_i as follows.

$$X_i = \begin{cases} 1 & \text{if the } i\text{th chosen marble is blue} \\ 0 & \text{otherwise} \end{cases}$$

   Then, we can write

$$X = X_1 + X_2 + \cdots + X_k$$

   Using the above equation, show

1. $EX = \frac{kb}{b+r}$.

2. $\text{Var}(X) = \frac{kbr}{(b+r)^2} \frac{b+r-k}{b+r-1}$.

Solution:

(a) We note that for any particular X_i, all marbles are equally likely to be chosen. This is because of symmetry: no marble is more likely to be chosen as the ith marble than any other marble. Therefore,

$$P(X_i = 1) = \frac{b}{b+r}, \qquad \text{for all } i \in \{1, 2, \cdots, k\}.$$

Therefore, $X_i \sim Bernoulli\left(\frac{b}{b+r}\right)$,

$$\begin{aligned} EX_i &= \frac{b}{b+r} \\ EX &= EX_1 + \cdots + EX_k \\ &= \frac{kb}{b+r} \end{aligned}$$

(b)

$$\begin{aligned}
\mathrm{Var}(X) &= \sum_{i=1}^{k} \mathrm{Var}(X_i) + 2\sum_{i<j} \mathrm{Cov}(X_i, X_j) \\
\mathrm{Var}(X_i) &= \frac{b}{b+r} \cdot \left(1 - \frac{b}{b+r}\right) \\
&= \frac{br}{(b+r)^2} \\
\mathrm{Cov}(X_i, X_j) &= E[X_i X_j] - E[X_i]E[X_j] \\
&= E[X_i X_j] - \left(\frac{b}{b+r}\right)^2 \\
E[X_i X_j] &= P(X_i = 1 \quad \& \quad X_j = 1) \\
&= P(X_1 = 1 \quad \& \quad X_2 = 1) \\
&= \frac{b}{b+r} \cdot \frac{b-1}{b+r-1} \\
\mathrm{Cov}(X_i, X_j) &= \frac{b(b-1)}{(b+r)(b+r-1)} - (\frac{b}{b+r})^2 \\
\mathrm{Var}(X) &= \frac{kbr}{(b+r)^2} + 2\binom{k}{2}\left[\frac{b(b-1)}{(b+r)(b+r-1)} - \left(\frac{b}{b+r}\right)^2\right] \\
&= \frac{kbr}{(b+r)^2} \cdot \frac{b+r-k}{b+r-1}
\end{aligned}$$

7. If $M_X(s) = \frac{1}{4} + \frac{1}{2}e^s + \frac{1}{4}e^{2s}$, find EX and $\mathrm{Var}(X)$.

Solution:

$$\begin{aligned}
M_X(s) &= \frac{1}{4} + \frac{1}{2}e^s + \frac{1}{4}e^{2s} \\
M'_X(s) &= \frac{1}{2}e^s + \frac{1}{2}e^{2s} \\
EX &= M'_X(0) \\
&= \frac{1}{2} + \frac{1}{2} \\
&= 1
\end{aligned}$$

$$\begin{aligned}
M''_X(s) &= \frac{1}{2}e^s + e^{2s} \\
EX^2 &= M''_X(0) \\
&= \frac{1}{2} + 1 \\
&= \frac{3}{2} \\
\text{Var}(x) &= \frac{3}{2} - 1 \\
&= \frac{1}{2}
\end{aligned}$$

9. (MGF of the Laplace distribution) Let X be a continuous random variable with the following PDF:

$$f_X(x) = \frac{\lambda}{2}e^{-\lambda|x|}.$$

Find the MGF of X, $M_X(s)$.

Solution:

$$\begin{aligned}
f_X(x) &= \frac{\lambda}{2}e^{-\lambda|x|} \\
M_X(s) &= E\left[e^{sX}\right] \\
&= \int_{-\infty}^{\infty} e^{sx} \cdot \frac{\lambda}{2}e^{-\lambda|x|}dx \\
&= \int_{-\infty}^{0} \frac{\lambda}{2}e^{(s+\lambda)x}dx + \int_{0}^{\infty} \frac{\lambda}{2}e^{(s-\lambda)x}dx \\
&= \left[\frac{\lambda}{2(s+\lambda)}e^{(s+\lambda)x}\right]_{-\infty}^{0} + \left[\frac{\lambda}{2(s-\lambda)}e^{(s-\lambda)x}\right]_{0}^{\infty} \\
&= \frac{\lambda}{2(s+\lambda)} + \frac{-\lambda}{2(s-\lambda)} \quad (\text{for} - \lambda < s < \lambda) \\
&= \frac{\lambda}{2}(\frac{1}{s+\lambda} + \frac{1}{\lambda - s}) \quad (\text{for} - \lambda < s < \lambda) \\
&= \frac{\lambda^2}{\lambda^2 - s^2} \quad (\text{for} - \lambda < s < \lambda)
\end{aligned}$$

11. Using the MGFs, show that if $Y = X_1 + X_2 + \cdots + X_n$, where X_i's are independent $Exponential(\lambda)$ random variables, then $Y \sim Gamma(n, \lambda)$.

Solution:

$$\begin{aligned}
X_i &\sim Exponential(\lambda) \\
M_{X_i}(s) &= \frac{\lambda}{\lambda - s} \quad (\text{for} \quad s < \lambda) \\
Y &= X_1 + \cdots + X_n \quad (X_i s \quad i.i.d.) \\
M_Y(s) &= (M_{X_1}(s))^n \\
&= \left(\frac{\lambda}{\lambda - s}\right)^n \\
&= MGF \quad of \quad Gamma(n, \lambda)
\end{aligned}$$

Therefore,

$$Y \sim Gamma(n, \lambda)$$

13. Let X and Y be two jointly continuous random variables with joint PDF

$$f_{X,Y}(x,y) = \begin{cases} \frac{1}{2}(3x + y) & 0 \leq x, y \leq 1 \\ \\ 0 & \text{otherwise} \end{cases}$$

and let the random vector $\mathbf{U}$ be defined as

$$\mathbf{U} = \begin{bmatrix} X \\ Y \end{bmatrix}.$$

(a) Find the mean vector of $\mathbf{U}$, $E\mathbf{U}$.

(b) Find the correlation matrix of $\mathbf{U}$, $\mathbf{R_U}$.

(c) Find the covariance matrix of $\mathbf{U}$, $\mathbf{C_U}$.

Solution:

$$\begin{aligned}
f_X(x) &= \int_0^1 \frac{1}{2}(3x+y)dy \\
&= \frac{3}{2}x + \frac{1}{4} \quad (\text{for} \quad 0 \leq x \leq 1) \\
f_Y(y) &= \int_0^1 \frac{1}{2}(3x+y)dx \\
&= \frac{3}{4} + \frac{y}{2} \quad (\text{for} \quad 0 \leq y \leq 1).
\end{aligned}$$

$$\begin{aligned}
EX &= \int_0^1 x\left(\frac{3}{2}x + \frac{1}{4}\right)dx \\
&= \frac{5}{8} \\
EX^2 &= \int_0^1 x^2\left(\frac{3}{2}x + \frac{1}{4}\right)dx \\
&= \frac{11}{24} \\
\text{Var}(X) &= \frac{11}{24} - \left(\frac{5}{8}\right)^2 \\
&= \frac{13}{192}.
\end{aligned}$$

$$\begin{aligned}
EY &= \int_0^1 \left(\frac{y^2}{2} + \frac{3}{4}y\right) dy \\
&= \frac{13}{24} \\
EY^2 &= \int_0^1 \left(\frac{y^3}{2} + \frac{3}{4}y^2\right) dy \\
&= \frac{3}{8} \\
\text{Var}(Y) &= \frac{3}{8} - \left(\frac{13}{24}\right)^2 \\
&= \frac{47}{576} \\
\text{Cov}(X,Y) &= EXY - EXEY \\
EXY &= \int_0^1 \int_0^1 \frac{xy}{2}(3x+y)dxdy \\
&= \frac{1}{3}
\end{aligned}$$

$$\begin{aligned}
\text{Cov}(X,Y) &= \frac{1}{3} - \frac{5}{8} \cdot \frac{13}{24} \\
&= \frac{-1}{192}
\end{aligned}$$

(a)

$$\begin{aligned}
E\mathbf{U} &= \begin{bmatrix} EX \\ EY \end{bmatrix} \\
&= \begin{bmatrix} \frac{5}{8} \\ \frac{13}{24} \end{bmatrix}
\end{aligned}$$

(b)

$$\begin{aligned}
\mathbf{R_U} &= \begin{bmatrix} EX^2 & EXY \\ EXY & EY^2 \end{bmatrix} \\
&= \begin{bmatrix} \frac{11}{24} & \frac{1}{3} \\ \frac{1}{3} & \frac{3}{8} \end{bmatrix}
\end{aligned}$$

(c)

$$\mathbf{C_U} = \begin{bmatrix} \mathrm{Var}(X) & \mathrm{Cov}(X,Y) \\ \mathrm{Cov}(X,Y) & \mathrm{Var}(Y) \end{bmatrix}$$
$$= \begin{bmatrix} \frac{13}{192} & \frac{-1}{192} \\ \frac{-1}{192} & \frac{47}{576} \end{bmatrix}$$

15. Let $\mathbf{X} = \begin{bmatrix} X_1 \\ X_2 \end{bmatrix}$ be a normal random vector with the following mean and covariance matrices:

$$\mathbf{m} = \begin{bmatrix} 1 \\ 2 \end{bmatrix}, \quad \mathbf{C} = \begin{bmatrix} 4 & 1 \\ 1 & 1 \end{bmatrix}.$$

Let also

$$\mathbf{A} = \begin{bmatrix} 2 & 1 \\ -1 & 1 \\ 1 & 3 \end{bmatrix}, \quad \mathbf{b} = \begin{bmatrix} -1 \\ 0 \\ 1 \end{bmatrix}, \quad \mathbf{Y} = \begin{bmatrix} Y_1 \\ Y_2 \\ Y_3 \end{bmatrix} = \mathbf{AX} + \mathbf{b}.$$

(a) Find $P(X_2 > 0)$.

(b) Find expected value vector of $\mathbf{Y}$, $\mathbf{m_Y} = E\mathbf{Y}$.

(c) Find the covariance matrix of $\mathbf{Y}$, $\mathbf{C_Y}$.

(d) Find $P(Y_2 \leq 2)$.

Solution:

$$X_1 \sim N(1, 4)$$
$$X_2 \sim N(2, 1)$$

(a)

$$\begin{aligned}
P(X_2 > 0) &= 1 - \Phi\left(\frac{0 - \mu_2}{\sigma_2}\right) \\
&= 1 - \Phi\left(\frac{-2}{1}\right) \\
&= 1 - \Phi(-2) \\
&= \Phi(2) \\
&\approx 0.98
\end{aligned}$$

(b)

$$\begin{aligned}
E\mathbf{Y} &= \mathbf{A}E\mathbf{X} + \mathbf{b} \\
&= \begin{bmatrix} 2 & 1 \\ -1 & 1 \\ 1 & 3 \end{bmatrix} \begin{bmatrix} 1 \\ 2 \end{bmatrix} + \begin{bmatrix} -1 \\ 0 \\ 1 \end{bmatrix} \\
&= \begin{bmatrix} 3 \\ 1 \\ 8 \end{bmatrix}
\end{aligned}$$

(c)

$$\begin{aligned}
\mathbf{C_Y} &= \mathbf{A}\mathbf{C_X}\mathbf{A}^T \\
&= \begin{bmatrix} 2 & 1 \\ -1 & 1 \\ 1 & 3 \end{bmatrix} \begin{bmatrix} 4 & 1 \\ 1 & 1 \end{bmatrix} \begin{bmatrix} 2 & -1 & 1 \\ 1 & 1 & 3 \end{bmatrix} \\
&= \begin{bmatrix} 21 & -6 & 18 \\ -6 & 3 & -3 \\ 18 & -3 & 19 \end{bmatrix}
\end{aligned}$$

(d)

$$\begin{aligned}
Y_2 &\sim N(1, 3) \\
P(Y_2 \leq 2) &= \Phi\left(\frac{2 - 1}{\sqrt{3}}\right) \\
&= \Phi\left(\frac{1}{\sqrt{3}}\right) \\
&\approx 0.718
\end{aligned}$$

17. A system consists of 4 components in a series, so the system works properly if all of the components are functional. In other words, the system fails if and only if at least one of its component fails. Suppose that we know that the probability that the component i fails is less than or equal to $p_f = \frac{1}{100}$, for $i = 1, 2, 3, 4$. Find an upper bound on the probability that the system fails.

Solution: Let F_i be the event that the ith component fails. Then,

$$\begin{aligned} P(F) &= P\left(\bigcup_{i=1}^{4} F_i\right) \\ &\leq \sum_{i=1}^{4} P(F_i) \\ &\leq \frac{4}{100} \end{aligned}$$

19. Let $X \sim Geometric(p)$. Using Markov's inequality, find an upper bound for $P(X \geq a)$, for a positive integer a. Compare the upper bound with the real value of $P(X \geq a)$.

Solution:

$$\begin{aligned} X &\sim Geometric(p) \\ EX &= \frac{1}{p}. \end{aligned}$$

$$\begin{aligned} &P(X \geq a) \\ &\leq \frac{EX}{a} \quad \text{(Using Markov's inequality)} \\ &= \frac{1}{pa} \end{aligned}$$

$$\begin{aligned}
P(X \geq a) &= \sum_{k=a}^{\infty} P(X = k) \\
&= \sum_{k=a}^{\infty} q^{k-1} p \\
&= pq^{a-1} \frac{1}{1-q} \\
&= q^{a-1} \\
&= (1-p)^{a-1}
\end{aligned}$$

We show $(1-p)^{a-1} \leq \frac{1}{pa}$ for all $a \geq 1$, $0 < p < 1$. To show this, look at the function:

$$\begin{aligned}
f(p) &= p(1-p)^{a-1} \\
f'(p) &= 0 \quad \text{which results in} \quad p = \frac{1}{a} \\
f(p) &\leq \frac{1}{a}(1 - \frac{1}{a})^{a-1} \leq \frac{1}{a} \\
p(1-p)^{a-1} &\leq \frac{1}{a} \\
(1-p)^{a-1} &\leq \frac{1}{pa}
\end{aligned}$$

21. (Cantelli's inequality) Let X be a random variable with $EX = 0$ and $\text{Var}(X) = \sigma^2$. We would like to prove that for any $a > 0$, we have

$$P(X \geq a) \leq \frac{\sigma^2}{\sigma^2 + a^2}.$$

This inequality is sometimes called the one-sided Chebyshev inequality.

Hint: One way to show this is to use $P(X \geq a) = P(X + c \geq a + c)$ for any constant $c \in \mathbb{R}$.

Solution:

$$\begin{aligned} P(X \geq a) &= P(X + c \geq a + c) \\ &= P\left((X+c)^2 \geq (a+c)^2\right) \\ &\leq \frac{E[(X+c)^2]}{(a+c)^2} \quad \text{(Markov's inequality)} \end{aligned}$$

We try to minimize $\frac{E[(X+c)^2]}{(a+c)^2}$ to get the best upper bound:

$$\begin{aligned} \frac{E[(X+c)^2]}{(a+c)^2} &= \frac{EX^2 + 2cEX + c^2}{(a+c)^2} \\ &= \frac{c^2 + \sigma^2}{(a+c)^2} \\ \frac{d}{dc} &= 0 \quad \text{.Thus,} \quad (2c)(a+c)^2 - 2(c+a)(c^2+\sigma^2) = 0 \\ c &= \frac{\sigma^2}{a} \quad \text{.Therefore,} \quad \frac{E[(X+c)^2]}{(a+c)^2} = \frac{\sigma^2}{\sigma^2 + a^2} \end{aligned}$$

23. Let X_i be i.i.d and $X_i \sim \textit{Exponential}(\lambda)$. Using Chernoff bounds, find an upper bound for $P(X_1 + X_2 + \cdots + X_n \geq an)$, where $a > \frac{1}{\lambda}$. Show that the bound goes to zero exponentially fast as a function of n.

Solution: Let $Y = X_1 + X_2 + \cdots + X_n$ then

$$\begin{aligned} M_Y(s) &= M_X(s)^n \\ &= \left(\frac{\lambda}{\lambda - s}\right)^n \quad (\text{for} \quad s < \lambda) \end{aligned}$$

Therefore,

$$\begin{aligned} P(Y \geq an) &\leq \min_{s>0} \left[e^{-san} M_Y(s)\right] \\ &= \min_{s>0} \left[e^{-san} \left(\frac{\lambda}{\lambda - s}\right)^n\right] \end{aligned}$$

$\frac{d}{ds} = 0$. Thus,

$$\begin{aligned}
-ane^{-san}\left(\frac{\lambda}{\lambda-s}\right)^n + \frac{n\lambda}{(\lambda-s)^2}\left(\frac{\lambda}{\lambda-s}\right)^{n-1} e^{-san} &= 0\\
-an + \frac{n}{\lambda-s} &= 0\\
s^* = \lambda - \frac{1}{a} &> 0 \quad (\text{since} \quad \lambda > \frac{1}{a})\\
P(Y \geq an) &\leq e^{-san}\left(\frac{\lambda}{\lambda-\lambda+\frac{1}{a}}\right)^n\\
&= e^{-san}(\lambda a)^n\\
&= \left[\lambda a e^{-(\lambda a-1)}\right]^n
\end{aligned}$$

Note that $xe^{-(x-1)} < 1$ for any $x > 1$. Since $\lambda a > 1$, we conclude $\lambda a e^{-(\lambda a-1)} < 1$. Therefore, $P(Y \geq an)$ goes to zero, exponentially fast in n.

25. Let X be a positive random variable with $EX = 10$. What can you say about the following quantities?

    (a) $E[X - X^3]$

    (b) $E[X \ln \sqrt{X}]$

    (c) $E\big[|2 - X|\big]$

Solution:

(a)

$$\begin{aligned}
g(X) &= X - X^3\\
g'(X) &= 1 - 3X^2\\
g''(X) &= -6X < 0 \quad (\text{for positive} \quad X).
\end{aligned}$$

Therefore, $g(X)$ is a concave function on $(0, \infty)$.

$$\begin{aligned} E[g(X)] &\leq g(E[X]) \\ E[X - X^3] &\leq \mu - \mu^3 \\ &= 10 - 1000 \\ &= -990 \end{aligned}$$

(b)

$$\begin{aligned} g(X) &= X \ln \sqrt{X} \\ &= \frac{1}{2} X \ln X \\ g'(X) &= \frac{1}{2} \ln X + \frac{1}{2} \\ g''(X) &= \frac{1}{2X} \quad (\text{for} \quad X > 0) \end{aligned}$$

$g(X)$ is a convex function on $(0, \infty)$. Thus,

$$\begin{aligned} E[g(X)] &\geq g(EX) \\ E[X \ln \sqrt{X}] &\geq \mu \ln \sqrt{\mu} \\ &= 10 \ln \sqrt{10} = 5 \ln 10 \end{aligned}$$

(c) Note that $|2 - X| = g(X)$ is a <u>convex</u> function on $(0, \infty)$.

$$\begin{aligned} E[|2 - X|] &\geq |2 - EX| \\ &= 8 \end{aligned}$$

Chapter 7

Limit Theorems and Convergence of RVs

1. Let X_i be i.i.d $Uniform(0,1)$. We define the sample mean as

$$M_n = \frac{X_1 + X_2 + ... + X_n}{n}.$$

(a) Find $E[M_n]$ and $\text{Var}(M_n)$ as a function of n.

(b) Using Chebyshev's inequality, find an upper bound on

$$P\left(\left|M_n - \frac{1}{2}\right| \geq \frac{1}{100}\right).$$

(c) Using your bound, show that

$$\lim_{n \to \infty} P\left(\left|M_n - \frac{1}{2}\right| \geq \frac{1}{100}\right) = 0.$$

Solution:

(a)

$$\begin{aligned}
EM_n &= \frac{EX_1 + \cdots + EX_n}{n} \\
&= \frac{nEX_1}{n} \\
&= EX_1 = \frac{1}{2} \\
\text{Var}(M_n) &= \frac{1}{n^2}\sum_{i=1}^{n} \text{Var}(X_i) \\
&= \frac{n\text{Var}X_1}{n^2} \\
&= \frac{\text{Var}(X_1)}{n} \\
&= \frac{\frac{1}{12}}{n} = \frac{1}{12n}
\end{aligned}$$

(b)

$$\begin{aligned}
P\left(\left|M_n - \frac{1}{2}\right| \geq \frac{1}{100}\right) &\leq \frac{\text{Var}(M_n)}{\left(\frac{1}{100}\right)^2} \\
&= \frac{10000}{12n}
\end{aligned}$$

(c)

$$\begin{aligned}
\lim_{n\to\infty} P\left(\left|M_n - \frac{1}{2}\right| \geq \frac{1}{100}\right) &\leq \lim_{n\to\infty} \frac{10000}{12n} = 0 \\
\lim_{n\to\infty} P\left(\left|M_n - \frac{1}{2}\right| \geq \frac{1}{100}\right) &= 0 \quad \text{(since probability is non-negative)}
\end{aligned}$$

3. In a communication system, each codeword consists of 1000 bits. Due to the noise, each bit may be received in error with probability 0.1. It is assumed bit errors occur independently. Since error correcting codes are used in this system, each codeword can be decoded reliably if there are fewer than or equal to 125 errors in the received codeword, otherwise the decoding fails.

Using the CLT, find the probability of decoding failure.

Solution:
Let $Y = X_1 + X_2 + \cdots + X_n, \quad n = 1000.$

$$\begin{aligned} X_i \sim &\quad \text{Bernoulli}(p = 0.1) \\ EX_i &= p = 0.1 \\ \text{Var}(X_i) &= p(1-p) = 0.09 \\ EY &= np = 100 \\ \text{Var}(Y) &= np(1-p) = 90 \end{aligned}$$

By the CLT:

$\frac{Y - EY}{\sqrt{\text{Var}(Y)}} = \frac{Y - 100}{\sqrt{90}}$ (can be approximated by $N(0,1)$). Thus,

$$\begin{aligned} P(Y > 125) &= P\left(\frac{Y-100}{\sqrt{90}} > \frac{125-100}{\sqrt{90}}\right) \\ &= 1 - \Phi\left(\frac{25}{\sqrt{90}}\right) \\ &\approx 0.0042 \end{aligned}$$

5. The amount of time needed for a certain machine to process a job is a random variable with mean $EX_i = 10$ minutes and $\text{Var}(X_i) = 2$ minutes^2. The time needed for different jobs are independent from each other. Find the probability that the machine processes fewer than or equal to 40 jobs in 7 hours.

   Solution: Let Y be the time that it takes to process 40 jobs. Then,

$$P(\text{Less than or equal to 40 jobs in 7 hours}) = P(Y > 7 \text{ hours}).$$

$$
\begin{aligned}
Y &= X_1 + X_2 + \cdots + X_{40} \\
EX_i &= 10, \text{Var}(X_i) = 2 \\
EY &= 40 \times 10 = 400 \\
\text{Var}(Y) &= 40 \times 2 = 80 \\
P(\text{Less than or equal to 40 jobs in 7 hours}) &= P(Y > 7 \times 60) \\
&= P(Y > 420) \\
&= P\left(\frac{Y - 400}{\sqrt{80}} > \frac{420 - 400}{\sqrt{80}}\right) \\
&\approx 1 - \Phi\left(\frac{20}{\sqrt{80}}\right) \approx 0.0127
\end{aligned}
$$

7. An engineer is measuring a quantity q. It is assumed that there is a random error in each measurement, so the engineer will take n measurements and report the average of the measurements as the estimated value of q. Specifically, if Y_i is the value that is obtained in the ith measurement, we assume that

$$Y_i = q + X_i,$$

where X_i is the error in the i'th measurement. We assume that X_i's are i.i.d with $EX_i = 0$ and $\text{Var}(X_i) = 4$ units. The engineer reports the average of measurements

$$M_n = \frac{Y_1 + Y_2 + ... + Y_n}{n}.$$

How many measurements does the engineer need to take until he is 95% sure that the final error is less than 0.1 units? In other words, what should the value of n be such that

$$P\big(q - 0.1 \leq M_n \leq q + 0.1\big) \geq 0.95 \ ?$$

Solution:

$$
\begin{aligned}
EY_i &= q + EX_i = q \\
\text{Var}(Y_i) &= \text{Var}(X_i) = 4 \\
Y &= Y_1 + \cdots + Y_n \quad \text{Thus:} \quad EY = nq \\
\text{Var}(Y) &= n\text{Var}(Y_i) = 4n.
\end{aligned}
$$

$$
\begin{aligned}
P(q-0.1 \leq M_n \leq q+0.1) &= P\left(q-0.1 \leq \frac{Y_1+\cdots+Y_n}{n} \leq q+0.1\right) \\
&= P(qn-0.1n \leq Y \leq qn+0.1n) \\
&= P\left(\frac{qn-0.1n-nq}{2\sqrt{n}} \leq \frac{Y-nq}{2\sqrt{n}} \leq \frac{qn+0.1n-nq}{2\sqrt{n}}\right) \\
&= P\left(-0.05\sqrt{n} \leq \frac{Y-nq}{2\sqrt{n}} \leq 0.05\sqrt{n}\right) \\
&\approx \Phi(0.05\sqrt{n}) - \Phi(-0.05\sqrt{n}) \\
&= 2\Phi\left(0.05\sqrt{n}\right) - 1 = 0.95
\end{aligned}
$$

Thus, we obtain:

$$
\begin{aligned}
\Phi\left(0.05\sqrt{n}\right) &= 0.975 \\
0.05\sqrt{n} &\geq 1.96 \\
n &\geq 1537
\end{aligned}
$$

9. Let X_2, X_3, X_4, $\cdots$ be a sequence of non-negative random variables such that

$$
F_{X_n}(x) = \begin{cases} \frac{e^{nx}+xe^{n}}{e^{nx}+\left(\frac{n+1}{n}\right)e^{n}} & 0 \leq x \leq 1 \\ \\ \frac{e^{nx}+e^{n}}{e^{nx}+\left(\frac{n+1}{n}\right)e^{n}} & x > 1 \end{cases}
$$

Show that X_n converges in distribution to $Uniform(0,1)$.

Solution: Since X_n's are non-negative we have

$$
F_{X_n}(x) = 0 \qquad \text{for } x < 0.
$$

For $0 < x < 1$,

$$\begin{aligned}\lim_{n\to\infty} F_{X_n}(x) &= \lim_{n\to\infty}\left[\frac{e^{nx}+xe^n}{e^{nx}+\left(\frac{n+1}{n}\right)e^n}\right]\\ &= \lim_{n\to\infty}\frac{xe^n}{\left(\frac{n+1}{n}\right)e^n}\\ &= \lim_{n\to\infty}\left(\frac{n}{n+1}\right)x\\ &= x\end{aligned}$$

For $x > 1$,

$$\begin{aligned}\lim_{F_{X_n}(x)\to\infty} &= \lim_{n\to\infty}\frac{e^{nx}}{e^{nx}}\\ &= 1\end{aligned}$$

$$\lim_{n\to\infty} F_{X_n}(x) = \begin{cases} 0 & x < 0 \\ 1 & x > 1 \\ x & 0 < x < 1 \end{cases}$$

$$X_n \xrightarrow{d} Uniform(0,1)$$

11. We perform the following random experiment. We put $n \geq 10$ blue balls and n red balls in a bag. We pick 10 balls at random (without replacement) from the bag. Let X_n be the number of blue balls chosen. We perform this experiment for $n = 10, 11, 12, \cdots$. Prove that $X_n \xrightarrow{d} Binomial\left(10, \frac{1}{2}\right)$.

Solution:

$$P(X_n = k) = \frac{\binom{n}{k}\cdot\binom{n}{10-k}}{\binom{2n}{10}} \quad \text{for} \quad k = 0, 1, 2, \cdots, 10$$

Note that for any fixed k, as n grows

$$\binom{n}{k} = \frac{n(n-1)\cdots(n-k+1)}{k!} \sim \frac{n^k}{k!}.$$

Using the above approximation:

$$P(X_n = k) \xrightarrow{as \quad n\to\infty} \frac{\frac{n^k}{k!}\frac{n^{10-k}}{(10-k)!}}{\frac{(2n)^{10}}{10!}}$$
$$= \frac{10!}{k!(10-k)!}\left(\frac{1}{2}\right)^{10}$$
$$= \binom{10}{k}\left(\frac{1}{2}\right)^{10}.$$

Thus,

$$\begin{cases} R_{X_n} = \{0, 1, 2, \cdots, 10\} \\ \lim_{n\to\infty} P(X_n = k) = \binom{10}{k}\left(\frac{1}{2}\right)^{10} \end{cases}$$

Therefore, using Theorem 7.1 in the text, we obtain

$$X_n \xrightarrow{d} Binomial(10, \frac{1}{2})$$

13. Let X_1, X_2, X_3, $\cdots$ be a sequence of continuous random variables such that

$$f_{X_n}(x) = \frac{n}{2}e^{-n|x|}.$$

Show that X_n converges in probability to 0.

Solution:

$$P\left(|X_n| > \epsilon\right) = 2\int_{\epsilon}^{\infty} f_{X_n}(x)dx \quad (\text{since} \quad f_{X_n}(-x) = f_{X_n}(x))$$
$$= 2\int_{\epsilon}^{\infty} \frac{n}{2}e^{-nx}dx$$
$$= \left[-e^{-nx}\right]_{\epsilon}^{\infty}$$
$$= e^{-n\epsilon}$$

Thus, $\lim\limits_{n\to\infty} P\left(|X_n| > \epsilon\right) = 0$

$$X_n \xrightarrow{p} 0$$

15. Let $Y_1, Y_2, Y_3, \cdots$ be a sequence of i.i.d random variables with mean $EY_i = \mu$ and finite variance $\text{Var}(Y_i) = \sigma^2$. Define the sequence $\{X_n, n = 2, 3, ...\}$ as

$$X_n = \frac{Y_1Y_2 + Y_2Y_3 + \cdots Y_{n-1}Y_n + Y_nY_1}{n}, \qquad \text{for } n = 2, 3, \cdots.$$

Show that $X_n \xrightarrow{p} \mu^2$.

Solution:

$$\begin{aligned} E[X_n] &= \frac{1}{n}\left[E\left[Y_1Y_2\right] + E\left[Y_2Y_3\right] + \cdots + E\left[Y_nY_1\right]\right] \\ &= \frac{1}{n} \cdot n \cdot EY_1 \cdot EY_2 \\ &= (\mu)^2. \end{aligned}$$

Also, for $n \geq 3$, we can write

$$\text{Var}(X_n) = \frac{1}{n^2}\left[n\text{Var}\left(Y_1Y_2\right) + 2n\text{Cov}\left(Y_1Y_2, Y_2Y_3\right)\right]$$

$$\begin{aligned} \text{Var}\left(Y_1Y_2\right) &= E\left[Y_1^2Y_2^2\right] - \left(E[Y_1Y_2]\right)^2 \\ &= E\left[Y_1\right]^2 E\left[Y_2\right]^2 - (\mu)^4 \\ &= \left(\sigma^2 + \mu^2\right)\left(\sigma^2 + \mu^2\right) - (\mu)^4 \\ &= \sigma^4 + 2(\mu^2)(\sigma^2) \end{aligned}$$

$$\begin{aligned} \text{Cov}\left(Y_1Y_2, Y_2Y_3\right) &= E\left[Y_1\right]E\left[Y_3\right]E\left[Y_2^2\right] - E\left[Y_1\right]E\left[Y_2\right]E\left[Y_2\right]E\left[Y_3\right] \\ &= \mu^2\left(\mu^2 + \sigma^2\right) - \left(\mu^4\right) \\ &= \mu^2\sigma^2 \end{aligned}$$

Therefore

$$\begin{aligned} \text{Var}(X_n) &= \frac{1}{n^2}\left[n\sigma^4 + 2n\mu^2\sigma^2 + 2n\mu^2\sigma^2\right] \\ &= \frac{1}{n}\left(\sigma^4 + 2\mu^2\sigma^2 + 2\mu^2\sigma^2\right) \end{aligned}$$

In particular $\quad \text{Var}(X_n) \to 0 \quad$ as $\quad n \to \infty$

Now, using Chebyshev's Inequality, we can write

$$P\left(|X_n - EX_n| > \epsilon\right) < \frac{\mathrm{Var}(X_n)}{\epsilon^2} \to 0 \quad \text{as} \quad n \to \infty$$
$$P\left(|X_n - EX_n| > \epsilon\right) \to 0 \quad \text{as} \quad n \to \infty.$$

Thus,

$$X_n \xrightarrow{p} \mu^2.$$

17. Let X_1, X_2, X_3, $\cdots$ be a sequence of random variables such that

$$X_n \sim Poisson(n\lambda), \qquad \text{for } n = 1, 2, 3, \cdots,$$

where $\lambda > 0$ is a constant. Define a new sequence Y_n as

$$Y_n = \frac{1}{n} X_n, \qquad \text{for } n = 1, 2, 3, \cdots.$$

Show that Y_n converges in mean square to λ, i.e., $Y_n \xrightarrow{m.s.} \lambda$.

Solution: Since $X_n \sim Poisson(n\lambda)$, we have

$$EX_n = n\lambda, \qquad \mathrm{Var}(X_n) = n\lambda.$$

$$EY_n = \frac{1}{n} EX_n = \frac{1}{n} \cdot n\lambda = \lambda.$$

We can write

$$\begin{aligned}
E[|Y_n - \lambda|^2] &= E\left[\left|\frac{1}{n} X_n - \lambda\right|^2\right] \\
&= \frac{1}{n^2} E[(X_n - n\lambda)^2] \\
&= \frac{1}{n^2} \mathrm{Var}(X_n) \\
&= \frac{1}{n^2} \cdot n\lambda = \frac{\lambda}{n} \to 0 \quad \text{as} \quad n \to \infty.
\end{aligned}$$

Thus, we conclude

$$Y_n \xrightarrow{m.s.} \lambda$$

19. Let X_1, X_2, X_3, $\cdots$ be a sequence of random variable such that $X_n \sim Rayleigh(\frac{1}{n})$, i.e.,

$$f_{X_n}(x) = \begin{cases} n^2 x e^{-\frac{n^2x^2}{2}} & x > 0 \\ 0 & \text{otherwise} \end{cases}$$

Show that $X_n \xrightarrow{a.s.} 0$.

Solution: Note that:

$$\begin{aligned} F_{X_n}(x) &= \int_0^x f_n(\alpha)d\alpha \\ &= 1 - e^{-\frac{n^2x^2}{2}} \\ \text{that} \quad P(|X_n| > \epsilon) &= P(X_n > \epsilon) \\ &= 1 - P(X_n < \epsilon) \\ &= e^{-\frac{n^2\epsilon^2}{2}}. \end{aligned}$$

Therefore,

$$\begin{aligned} \sum_{n=1}^{\infty} P(|X_n| > \epsilon) &= \sum_{n=1}^{\infty} e^{-\frac{n^2\epsilon^2}{2}} \\ &\leq \sum_{n=1}^{\infty} e^{-\frac{n\epsilon^2}{2}} \\ &= \frac{e^{-\frac{\epsilon^2}{2}}}{1 - e^{-\frac{\epsilon^2}{2}}} < \infty. \end{aligned}$$

Therefore, using Theorem 7.5, we conclude

$$X_n \xrightarrow{a.s.} 0$$

Chapter 8

Statistical Inference I: Classical Methods

1. Let X be the weight of a randomly chosen individual from a population of adult men. In order to estimate the mean and variance of X, we observe a random sample $X_1, X_2, \cdots, X_{10}$. Thus, the X_i's are i.i.d. and have the same distribution as X. We obtain the following values (in pounds):

   165.5, 175.4, 144.1, 178.5, 168.0, 157.9, 170.1, 202.5, 145.5, 135.7

   Find the values of the sample mean, the sample variance, and the sample standard deviation for the observed sample.

   Solution: The sample mean is

   $$\begin{aligned}\overline{X} &= \frac{X_1 + X_2 + X_3 + X_4 + X_5 + X_6 + X_7 + X_8 + X_9 + X_{10}}{10} \\ &= (165.5 + 175.4 + 144.1 + 178.5 + 168.0 + 157.9 + 170.1 + 202.5 + \\ &\quad 145.5 + 135.7)/10 \\ &= 164.32\end{aligned}$$

   The sample variance is given by

   $$S^2 = \frac{1}{10-1}\sum_{k=1}^{10}(X_k - 164.32)^2 = 383.70,$$

   and the sample standard deviation is given by

   $$S = \sqrt{S^2} = 19.59.$$

You can use the following MATLAB code to compute the above values:

```
x=[165.5, 175.4, 144.1, 178.5, 168.0, 157.9, 170.1,
202.5, 145.5, 135.7];
m=mean(x);
v=var(x);
s=std(x);
```

3. Let X_1, X_2, X_3, ..., X_n be a random sample from the following distribution:

$$f_X(x) = \begin{cases} \theta\left(x - \frac{1}{2}\right) + 1 & \text{for } 0 \leq x \leq 1 \\ 0 & \text{otherwise} \end{cases}$$

where $\theta \in [-2, 2]$ is an unknown parameter. We define the estimator $\hat{\Theta}_n$ as

$$\hat{\Theta}_n = 12\overline{X} - 6$$

to estimate θ.

(a) Is $\hat{\Theta}_n$ an unbiased estimator of θ?

(b) Is $\hat{\Theta}_n$ a consistent estimator of θ?

(c) Find the mean squared error (MSE) of $\hat{\Theta}_n$.

Solution: Let's first EX and $\textrm{Var}(X)$ in terms of θ. We have

$$\begin{aligned} EX &= \int_0^1 x\left[\theta\left(x - \frac{1}{2}\right) + 1\right] dx \\ &= \frac{\theta + 6}{12}, \end{aligned}$$

$$
\begin{aligned}
EX^2 &= \int_0^1 x^2 \left[\theta\left(x - \frac{1}{2}\right) + 1\right] \, dx \\
&= \frac{\theta + 4}{12},
\end{aligned}
$$

$$
\begin{aligned}
\text{Var}(X) &= EX^2 - EX^2 \\
&= \frac{12 - \theta^2}{144}.
\end{aligned}
$$

(a) Is $\hat{\Theta}_n$ an unbiased estimator of θ? To see this, we write

$$
\begin{aligned}
E[\hat{\Theta}_n] &= E[12\overline{X} - 6] \\
&= 12E[\overline{X}] - 6 \\
&= 12 \cdot \frac{\theta + 6}{12} - 6 \\
&= \theta.
\end{aligned}
$$

Thus, $\hat{\Theta}_n$ IS an unbiased estimator of θ.

(b) To show that $\hat{\Theta}_n$ is a consistent estimator of θ, we need to show

$$
\lim_{n \to \infty} P\big(|\hat{\Theta}_n - \theta| \geq \epsilon\big) = 0, \qquad \text{for all } \epsilon > 0.
$$

Since $\hat{\Theta}_n = 12\overline{X} - 6$ and $\theta = 12EX - 6$, we conclude

$$
\begin{aligned}
P\big(|\hat{\Theta}_n - \theta| \geq \epsilon\big) &= P\big(12|\overline{X} - EX| \geq \epsilon\big) \\
&= P\big(|\overline{X} - EX| \geq \frac{\epsilon}{12}\big)
\end{aligned}
$$

which goes to zero as $n \to \infty$ by the law of large numbers. Therefore, $\hat{\Theta}_n$ is a consistent estimator of θ.

(c) To find the mean squared error (MSE) of $\hat{\Theta}_n$, we write

$$\begin{aligned} MSE(\hat{\Theta}_n) &= \text{Var}(\hat{\Theta}_n) + B(\hat{\Theta}_n)^2 \\ &= \text{Var}(\hat{\Theta}_n) \\ &= \text{Var}(12\overline{X} - 6) \\ &= 144\text{Var}(\overline{X}) \\ &= 144\frac{\text{Var}(X)}{n} \\ &= 144 \cdot \frac{12 - \theta^2}{144n} \\ &= \frac{12 - \theta^2}{n}. \end{aligned}$$

Note that this gives us another way to argue that $\hat{\Theta}_n$ is a consistent estimator of θ. In particular, since

$$\lim_{n \to \infty} MSE(\hat{\Theta}_n) = 0,$$

we conclude that $\hat{\Theta}_n$ is a consistent estimator of θ.

5. Let $X_1, \ldots, X_4$ be a random sample from an *Exponential*(θ) distribution. Suppose we observed $(x_1, x_2, x_3, x_4) = (2.35, 1.55, 3.25, 2.65)$. Find the likelihood function using

$$f_{X_i}(x_i; \theta) = \theta e^{-\theta x_i}, \qquad \text{for } x_i \geq 0$$

as the PDF.

Solution: If $X_i \sim Exponential(\theta)$, then

$$f_{X_i}(x; \theta) = \theta e^{-\theta x}$$

Thus, for $x_i \geq 0$, we can write

$$\begin{aligned} L(x_1, x_2, x_3, x_4; \theta) &= f_{X_1 X_2 X_3 X_4}(x_1, x_2, x_3, x_4; \theta) \\ &= f_{X_1}(x_1; \theta) f_{X_2}(x_2; \theta) f_{X_3}(x_3; \theta) f_{X_4}(x_4; \theta) \\ &= \theta^4 e^{-(x_1 + x_2 + x_3 + x_4)\theta}. \end{aligned}$$

Since we have observed $(x_1, x_2, x_3, x_4) = (2.35, 1.55, 3.25, 2.65)$, we have

$$L(2.35, 1.55, 3.25, 2.65; \theta) = \theta^4 e^{-9.8\theta}.$$

7. Let X be one observation from a $N(0, \sigma^2)$ distribution.

   (a) Find an unbiased estimator of σ^2.

   (b) Find the log likelihood, $\log(L(x; \sigma^2))$, using

   $$f_X(x; \sigma^2) = \frac{1}{\sqrt{2\pi}\sigma} exp\left\{-\frac{x^2}{2\sigma^2}\right\}$$

   as the PDF.

   (c) Find the Maximum Likelihood Estimate (MLE) for the standard deviation σ, $\hat{\sigma}_{ML}$.

Solution:

(a) Note that

$$E(X^2) = \text{Var}(X) + (EX)^2 = \sigma^2 + \mu^2 = \sigma^2.$$

Therefore $\hat{\sigma}(X) = X^2$ is an unbiased estimator of σ^2.

(b) The likelihood function is

$$L(x; \sigma^2) = f_X(x; \sigma^2) = \frac{1}{\sqrt{2\pi}\sigma} e^{-\frac{1}{2\sigma^2}(x)^2}.$$

The log-likelihood function is

$$\ln L(x; \sigma^2) = -\ln(2\pi)^{\frac{1}{2}} - \ln\sigma - \frac{x^2}{2\sigma^2}.$$

(c) To find the MLE for σ, we differentiate $\ln L(x;\sigma^2)$ with respect to σ and set it equal to zero.

$$\begin{aligned}\frac{\partial}{\partial\sigma}\ln L &= -\frac{1}{\sigma}+\frac{x^2}{\sigma^3}\\ &= -\frac{1}{\sigma}+\frac{x^2}{\sigma^3} \overset{set}{=} 0.\end{aligned}$$

Therefore,

$$\hat{\sigma}X^2 = \hat{\sigma}^3 \rightarrow \hat{\sigma} = |X|.$$

Also, we can verify that the second derivative is negative to make sure that $\hat{\sigma} = |X|$ is actually the maximizing value:

$$\frac{\partial^2}{\partial\sigma^2}\ln L = \frac{1}{\sigma^2} - \frac{3x^2}{\sigma^4} < 0 \text{ when } \hat{\sigma} = |x|.$$

9. In this problem, we would like to find the CDFs of the order statistics. Let $X_1, \ldots, X_n$ be a random sample from a continuous distribution with CDF $F_X(x)$ and PDF $f_X(x)$. Define $X_{(1)}, \ldots, X_{(n)}$ as the order statistics and show that

$$F_{X_{(i)}}(x) = \sum_{k=i}^{n}\binom{n}{k}\big[F_X(x)\big]^k\big[1-F_X(x)\big]^{n-k}.$$

Hint: Fix $x \in \mathbb{R}$. Let Y be a random variable that counts the number of $X_j's \leq x$. Define $\{X_j \leq x\}$ as a "success" and $\{X_j > x\}$ as a "failure", and show that $Y \sim Binomial(n, p = F_X(x))$.

Solution:

Let Y be a random variable thats counts the number of $X_1, \ldots, X_n \leq x$ where x is fixed. Now if we define $\{X_j \leq x\}$ as a "success," $Y \sim binomial(n, F_X(x))$. The event $\{X_{(i)} \leq x\}$ is equivalent to the event $\{Y \geq i\}$, so

$$F_{X_{(i)}}(x) = P(Y \geq i) = \sum_{k=i}^{n}\binom{n}{k}\big[F_X(x)\big]^k\big[1-F_X(x)\big]^{n-k}.$$

11. A random sample X_1, X_2, X_3, ..., X_{100} is given from a distribution with known variance $\text{Var}(X_i) = 81$. For the observed sample, the sample mean is $\overline{X} = 50.1$. Find an approximate 95% confidence interval for $\theta = EX_i$.

    Solution: Since n is large, a 95% CI can be expressed as given by

$$\left[\overline{X} - z_{0.025}\sqrt{\frac{\text{Var}(X_i)}{n}}, \overline{X} + z_{0.025}\sqrt{\frac{\text{Var}(X_i)}{n}}\right].$$

    If we plug in known values, the 95% CI is (48.3, 51.9).

13. Let X_1, X_2, X_3, ..., X_{100} be a random sample from a distribution with unknown variance $\text{Var}(X_i) = \sigma^2 < \infty$. For the observed sample, the sample mean is $\overline{X} = 110.5$, and the sample variance is $S^2 = 45.6$. Find a 95% confidence interval for $\theta = EX_i$.

    Solution: Since n is relatively large, the interval

$$\left[\overline{X} - z_{\frac{\alpha}{2}}\frac{S}{\sqrt{n}}, \overline{X} + z_{\frac{\alpha}{2}}\frac{S}{\sqrt{n}}\right]$$

    is approximately a $(1-\alpha)100\%$ confidence interval for θ. Here, $n = 100$, $\alpha = .05$, so we need

$$z_{\frac{\alpha}{2}} = z_{0.025} = \Phi^{-1}(1 - 0.025) = 1.96.$$

    Thus, we can obtain a 95% confidence interval for μ as

$$\begin{aligned}\left[\overline{X} - z_{\frac{\alpha}{2}}\frac{S}{\sqrt{n}}, \overline{X} + z_{\frac{\alpha}{2}}\frac{S}{\sqrt{n}}\right] &= \left[110.5 - 1.96 \cdot \frac{\sqrt{45.6}}{10}, 110.5 + 1.96 \cdot \frac{\sqrt{45.6}}{10}\right] \\ &\approx [109.18, 111.82]\end{aligned}$$

    Therefore, $[109.18, 111.82]$ is an approximate 95% confidence interval for μ.

15. Let X_1, X_2, X_3, X_4, X_5 be a random sample from a $N(\mu, 1)$ distribution, where μ is unknown. Suppose that we have observed the following values

$$5.45, \quad 4.23, \quad 7.22, \quad 6.94, \quad 5.98$$

We would like to decide between

$$H_0\colon \mu = \mu_0 = 5,$$
$$H_1\colon \mu \neq 5.$$

(a) Define a test statistic to test the hypotheses and draw a conclusion assuming $\alpha = 0.05$.

(b) Find a 95% confidence interval around $\overline{X}$. Is μ_0 included in the interval? How does the exclusion of μ_0 in the interval relate to the hypotheses we are testing?

Solution:

(a) Here we define the test statistic as

$$\begin{aligned} W &= \frac{\overline{X} - \mu_0}{\sigma/\sqrt{n}} \\ &= \frac{5.96 - 5}{1/\sqrt{5}} \\ &\approx 2.15. \end{aligned}$$

Here, $\alpha = .05$, so $z_{\frac{\alpha}{2}} = z_{0.025} = 1.96$. Since $|W| > z_{\frac{\alpha}{2}}$, we reject H_0 and accept H_1.

(b) The 95% CI is given by

$$\left(5.96 - 1.96 * \frac{1}{\sqrt{(5)}}, 5.96 + 1.96 * \frac{1}{\sqrt{(5)}}\right) = (5.09, 6.84).$$

Since μ_0 is not included in the interval, we are able to reject the null hypothesis and conclude that μ is not 5.

17. Let X_1, X_2 ,..., X_{150} be a random sample from an unknown distribution. After observing this sample, the sample mean and the sample variance are calculated to be as follows:

$$\overline{X} = 52.28, \qquad S^2 = 30.9$$

Design a level 0.05 test to choose between

H_0: $\mu = 50$,

H_1: $\mu > 50$.

Do you accept or reject H_0?

Solution:

$$\begin{aligned} W &= \frac{\overline{X} - \mu_0}{S/\sqrt{n}} \\ &= \frac{52.28 - 50}{\sqrt{30.9/150}} \\ &= 5.03 \end{aligned}$$

Since $5.03 > 1.96$, we reject H_0.

19. Let X_1, X_2 ,..., X_{121} be a random sample from an unknown distribution. After observing this sample, the sample mean and the sample variance are calculated to be as follows:

$$\overline{X} = 29.25, \qquad S^2 = 20.7$$

Design a test to decide between

H_0: $\mu = 30$,

H_1: $\mu < 30$,

and calculate the P-value for the observed data.

Solution: We define the test statistic as

$$\begin{aligned} W &= \frac{\overline{X} - \mu_0}{S/\sqrt{n}} \\ &= \frac{29.25 - 30}{\sqrt{20.7}/\sqrt{121}} \\ &= -1.81 \end{aligned}$$

and by Table 8.4 the test threshold is $-z_\alpha$. The P-value is $P(\text{type I error})$ when the test threshold c is chosen to be $c = -1.81$. Thus,

$$-z_\alpha = 1.81$$

Noting that by definition $z_\alpha = \Phi^{-1}(1 - \alpha)$, we obtain $P(\text{type I error})$ as

$$\alpha = 1 - \Phi(1.81) \approx 0.035$$

Therefore,

$$P - \text{value} = 0.035$$

21. Consider the following observed values of (x_i, y_i):

$$(-5, -2), \quad (-3, 1), \quad (0, 4), \quad (2, 6), \quad (1, 3).$$

   (a) Find the estimated regression line

$$\hat{y} = \hat{\beta}_0 + \hat{\beta}_1 x$$

   based on the observed data.

   (b) For each x_i, compute the fitted value of y_i using

$$\hat{y}_i = \hat{\beta}_0 + \hat{\beta}_1 x_i.$$

   (c) Compute the residuals, $e_i = y_i - \hat{y}_i$.

   (d) Calculate R-squared.

Solution:

(a) We have

$$\begin{aligned}
\overline{x} &= \frac{-5-3+0+2+1}{5} = -1 \\
\overline{y} &= \frac{-2+1+4+6+3}{5} = 2.4 \\
s_{xx} &= (-5+1)^2 + (-3+1)^2 + (0+1)^2 + (2+1)^2 + (1+1)^2 = 34 \\
s_{xy} &= (-5+1)(-2-2.4) + (-3+1)(1-2.4) + (0+1)(4-2.4) \\
&\quad + (2+1)(6-2.4) + (1+1)(3-2.4) = 34.
\end{aligned}$$

Therefore, we obtain

$$\begin{aligned}
\hat{\beta}_1 &= \frac{s_{xy}}{s_{xx}} = \frac{34}{34} = 1 \\
\hat{\beta}_0 &= 2.4 - (1)(-1) = 3.4.
\end{aligned}$$

(b) The fitted values are given by

$$\hat{y}_i = 3.4 + 1x_i,$$

so we obtain

$$\hat{y}_1 = -1.6, \qquad \hat{y}_2 = 0.4, \qquad \hat{y}_3 = 3.4, \qquad \hat{y}_4 = 5.4, \qquad \hat{y}_4 = 4.4.$$

(c) We have

$$\begin{aligned}
e_1 &= y_1 - \hat{y}_1 = -2 + 1.6 = -0.4, \\
e_2 &= y_2 - \hat{y}_2 = 1 - 0.4 = 0.6, \\
e_3 &= y_3 - \hat{y}_3 = 4 - 3.4 = 0.6, \\
e_4 &= y_4 - \hat{y}_4 = 6 - 5.4 = 0.6 \\
e_4 &= y_4 - \hat{y}_4 = 3 - 4.4 = -1.4.
\end{aligned}$$

(d) We have

$$\begin{aligned}
s_{yy} &= (-2-2.4)^2 + (1-2.4)^2 + (4-2.4)^2 + (6-2.4)^2 + (3-2.4)^2 \\
&= 37.2.
\end{aligned}$$

We conclude

$$r^2 = \frac{(34)^2}{34 \times 37.2} \approx 0.914.$$

23. Consider the simple linear regression model

$$Y_i = \beta_0 + \beta_1 x_i + \epsilon_i,$$

where ϵ_i's are independent $N(0, \sigma^2)$ random variables. Therefore, Y_i is a normal random variable with mean $\beta_0 + \beta_1 x_i$ and variance σ^2. Moreover, Y_i's are independent. As usual, we have the observed data pairs (x_1, y_1), (x_2, y_2), $\cdots$, (x_n, y_n) from which we would like to estimate β_0 and β_1. In this chapter, we found the following estimators

$$\hat{\beta}_1 = \frac{s_{xy}}{s_{xx}},$$
$$\hat{\beta}_0 = \overline{Y} - \hat{\beta}_1 \overline{x}.$$

where

$$s_{xx} = \sum_{i=1}^{n} (x_i - \overline{x})^2,$$
$$s_{xy} = \sum_{i=1}^{n} (x_i - \overline{x})(Y_i - \overline{Y}).$$

(a) Show that $\hat{\beta}_1$ is a normal random variable.

(b) Show that $\hat{\beta}_1$ is an unbiased estimator of β_1, i.e.,

$$E[\hat{\beta}_1] = \beta_1.$$

(c) Show that

$$\mathrm{Var}(\hat{\beta}_1) = \frac{\sigma^2}{s_{xx}}.$$

Solution:

(a) Note that

$$\begin{aligned}\hat{\beta}_1 &= \frac{s_{xy}}{s_{xx}} \\ &= \frac{\sum_{i=1}^{n}(x_i - \overline{x})(Y_i - \overline{Y})}{s_{xx}} \\ &= \frac{\sum_{i=1}^{n}(x_i - \overline{x})Y_i}{s_{xx}} - \frac{\overline{Y}\sum_{i=1}^{n}(x_i - \overline{x})}{s_{xx}} \\ &= \frac{\sum_{i=1}^{n}(x_i - \overline{x})Y_i}{s_{xx}}.\end{aligned}$$

Thus, $\hat{\beta}_1$ can be written as a linear combination of Y_i's, i.e.,

$$\hat{\beta}_1 = \sum_{i=1}^{n} c_i Y_i.$$

Since the Y_i's are normal and independent, we conclude that $\hat{\beta}_1$ is a normal random variable.

(b) Note that

$$\begin{aligned}Y_i - \overline{Y} &= (\beta_0 + \beta_1 x_i + \epsilon_i) - (\beta_0 + \beta_1 \overline{x} + \overline{\epsilon}) \\ &= \beta_1(x_i - \overline{x}) + (\epsilon_i - \overline{\epsilon}).\end{aligned}$$

Therefore,

$$\begin{aligned}E[Y_i - \overline{Y}] &= \beta_1(x_i - \overline{x}) + E[\epsilon_i - \overline{\epsilon}] \\ &= \beta_1(x_i - \overline{x}).\end{aligned}$$

Thus,

$$\begin{aligned}E[\hat{\beta}_1] &= \frac{\sum_{i=1}^{n}(x_i - \overline{x})E[Y_i - \overline{Y}]}{s_{xx}} \\ &= \frac{\sum_{i=1}^{n}(x_i - \overline{x})\beta_1(x_i - \overline{x})}{s_{xx}} \\ &= \beta_1.\end{aligned}$$

(c) We have

$$\hat{\beta}_1 = \frac{\sum_{i=1}^{n}(x_i - \overline{x})Y_i}{s_{xx}},$$

where the Y_i's are independent, so

$$\begin{aligned}
\mathrm{Var}(\hat{\beta}_1) &= \frac{\sum_{i=1}^{n}(x_i - \overline{x})^2 \mathrm{Var}(Y_i)}{s_{xx}^2} \\
&= \frac{\sum_{i=1}^{n}(x_i - \overline{x})^2 \sigma^2}{s_{xx}^2} \\
&= \frac{\sigma^2}{s_{xx}}.
\end{aligned}$$

Chapter 9

Statistical Inference II: Bayesian Inference

1. Let X be a continuous random variable with the following PDF

$$f_X(x) = \begin{cases} 6x(1-x) & \text{if } 0 \leq x \leq 1 \\ \\ 0 & \text{otherwise} \end{cases}$$

Suppose that we know

$$Y \mid X = x \quad \sim \quad Geometric(x).$$

Find the posterior density of X given $Y = 2$, $f_{X|Y}(x|2)$.

Solution: Using Bayes' rule, we have

$$f_{X|Y}(x|2) = \frac{P_{Y|X}(2|x) f_X(x)}{P_Y(2)}.$$

We know $Y \mid X = x \quad \sim \quad Geometric(x)$, so

$$P_{Y|X}(y|x) = x(1-x)^{y-1}, \qquad \text{for } y = 1, 2, \cdots.$$

Therefore,

$$P_{Y|X}(2|x) = x(1-x).$$

To find $P_Y(2)$, we can use the law of total probability

$$\begin{aligned} P_Y(2) &= \int_{-\infty}^{\infty} P_{Y|X}(2|x) f_X(x) \quad dx \\ &= \int_0^1 x(1-x) \cdot 6x(1-x) \quad dx \\ &= \frac{1}{5}. \end{aligned}$$

Therefore, we obtain

$$\begin{aligned} f_{X|Y}(x|2) &= \frac{6x^2(1-x)^2}{\frac{1}{5}} \\ &= 30x^2(1-x)^2, \qquad \text{for } 0 \leq x \leq 1. \end{aligned}$$

3. Let X and Y be two jointly continuous random variables with joint PDF

$$f_{XY}(x,y) = \begin{cases} x + \frac{3}{2}y^2 & 0 \leq x, y \leq 1 \\ 0 & \text{otherwise.} \end{cases}$$

Find the MAP and the ML estimates of X given $Y = y$.

Solution: For $0 \leq x \leq 1$, we have

$$\begin{aligned} f_X(x) &= \int_{-\infty}^{\infty} f_{XY}(x,y) dy \\ &= \int_0^1 \left(x + \frac{3}{2}y^2 \right) dy \\ &= \left[xy + \frac{1}{2}y^3 \right]_0^1 \\ &= x + \frac{1}{2}. \end{aligned}$$

Thus,

$$f_X(x) = \begin{cases} x + \frac{1}{2} & 0 \leq x \leq 1 \\ \\ 0 & \text{otherwise} \end{cases}$$

Similarly, for $0 \leq y \leq 1$, we have

$$\begin{aligned} f_Y(y) &= \int_{-\infty}^{\infty} f_{XY}(x, y) dx \\ &= \int_0^1 \left(x + \frac{3}{2} y^2 \right) dx \\ &= \left[\frac{1}{2} x^2 + \frac{3}{2} y^2 x \right]_0^1 \\ &= \frac{3}{2} y^2 + \frac{1}{2}. \end{aligned}$$

Thus,

$$f_Y(y) = \begin{cases} \frac{3}{2} y^2 + \frac{1}{2} & 0 \leq y \leq 1 \\ \\ 0 & \text{otherwise} \end{cases}$$

The MAP estimate of X, given $Y = y$, is the value of x that maximizes

$$f_{X|Y}(x|y) = \frac{x + \frac{3}{2} y^2}{\frac{3}{2} y^2 + \frac{1}{2}}, \qquad \text{for } 0 \leq x, y \leq 1.$$

For any $y \in [0, 1]$, the above function is maximized at $x = 1$. Thus, we obtain the MAP estimate of x as

$$\hat{x}_{MAP} = 1.$$

The ML estimate of X, given $Y = y$, is the value of x that maximizes

$$\begin{aligned} f_{Y|X}(y|x) &= \frac{x + \frac{3}{2} y^2}{x + \frac{1}{2}} \\ &= 1 + \frac{\frac{3}{2} y^2 - \frac{1}{2}}{x + \frac{1}{2}}, \qquad \text{for } 0 \leq x, y \leq 1. \end{aligned}$$

Therefore, we conclude

$$\hat{x}_{ML} = \begin{cases} 1 & 0 \leq y \leq \frac{1}{\sqrt{3}} \\ \\ 0 & \text{otherwise} \end{cases}$$

5. Let $X \sim N(0,1)$ and

$$Y = 2X + W,$$

where $W \sim N(0,1)$ is independent of X.

(a) Find the MMSE estimator of X given Y, ($\hat{X}_M$).

(b) Find the MSE of this estimator, using $MSE = E[(X - \hat{X_M})^2]$.

(c) Check that $E[X]^2 = E[\hat{X}_M^2] + E[\tilde{X}^2]$.

Solution: Since X and W are independent and normal, Y is also normal. Moreover, X and Y are jointly normal.

$$\begin{aligned}\text{Cov}(X,Y) &= \text{Cov}(X, 2X + W) \\ &= 2\text{Cov}(X,X) + \text{Cov}(X,W) \\ &= 2\text{Var}(X) = 2.\end{aligned}$$

Therefore,

$$\begin{aligned}\rho(X,Y) &= \frac{\text{Cov}(X,Y)}{\sigma_X \sigma_Y} \\ &= \frac{2}{1 \cdot \sqrt{5}} = \frac{2}{\sqrt{5}}.\end{aligned}$$

(a) The MMSE estimator of X given Y is

$$\begin{aligned}\hat{X}_M &= E[X|Y] \\ &= \mu_X + \rho\sigma_X \frac{Y - \mu_Y}{\sigma_Y} \\ &= \frac{2Y}{5}.\end{aligned}$$

(b) The MSE of this estimator is given by

$$\begin{aligned}
E[(X-\hat{X}_M)^2] &= E\left[\left(X-\frac{2Y}{5}\right)^2\right]\\
&= E\left[\left(X-\frac{4}{5}X-\frac{2}{5}W\right)^2\right]\\
&= E\left[\left(\frac{1}{5}X-\frac{2}{5}W\right)^2\right]\\
&= \frac{1}{25}E\left[(X-2W)^2\right]\\
&= \frac{1}{25}[EX^2+4EW^2]\\
&= \frac{1}{5}.
\end{aligned}$$

(c) Note that $E[X]^2 = 1$. Also,

$$E[\hat{X}_M^2] = \frac{4EY^2}{25} = \frac{4}{5}.$$

In the above, we also found, $MSE = E[\tilde{X}^2] = \frac{1}{5}$. Therefore, we have

$$E[X]^2 = E[\hat{X}_M^2] + E[\tilde{X}^2].$$

7. Suppose that the signal $X \sim N(0, \sigma_X^2)$ is transmitted over a communication channel. Assume that the received signal is given by

$$Y = X + W,$$

where $W \sim N(0, \sigma_W^2)$ is independent of X.

(a) Find the MMSE estimator of X given Y, $(\hat{X}_M)$.

(b) Find the MSE of this estimator.

Solution: Since X and W are independent and normal, Y is also normal. The variance is

$$\begin{aligned}\mathrm{Cov}(X,Y) &= \mathrm{Cov}(X, X+W)\\ &= \mathrm{Cov}(X) + \mathrm{Cov}(X,W)\\ &= \mathrm{Var}(X) = \sigma_X^2.\end{aligned}$$

Therefore,

$$\begin{aligned}\rho(X,Y) &= \frac{\mathrm{Cov}(X,Y)}{\sigma_X \sigma_Y}\\ &= \frac{\sigma_X}{\sqrt{\sigma_X^2+\sigma_W^2}}.\end{aligned}$$

(a) The MMSE estimator of X given Y is

$$\begin{aligned}\hat{X}_M &= E[X|Y]\\ &= \mu_X + \rho\sigma_X \frac{Y-\mu_Y}{\sigma_Y}\\ &= \frac{\sigma_X^2}{\sigma_X^2+\sigma_W^2} Y.\end{aligned}$$

(b) The MSE of this estimator is given by

$$\begin{aligned}E[(X-\hat{X_M})^2] &= E\left[\tilde{X}^2\right]\\ &= E[X^2] - E[\hat{X_M}^2]\\ &= \sigma_X^2 - \left(\frac{\sigma_X^2}{\sigma_X^2+\sigma_W^2}\right)^2 (\sigma_X^2+\sigma_W^2)\\ &= \frac{\sigma_X^2\sigma_W^2}{\sigma_X^2+\sigma_W^2}.\end{aligned}$$

9. Consider again Problem 8, in which X is an unobserved random variable with $EX=0$, $\mathrm{Var}(X)=5$. Assume that we have observed Y_1 and Y_2 given

by

$$Y_1 = 2X + W_1,$$
$$Y_2 = X + W_2,$$

where $EW_1 = EW_2 = 0$, $\text{Var}(W_1) = 2$, and $\text{Var}(W_2) = 5$. Assume that W_1, W_2 , and X are independent random variables. Find the linear MMSE estimator of X, given Y_1 and Y_2 using the vector formula

$$\hat{\mathbf{X}}_L = \mathbf{C_{XY}}\mathbf{C_Y}^{-1}(\mathbf{Y} - E[\mathbf{Y}]) + E[\mathbf{X}].$$

Solution: Note that here X is a one dimensional vector, and $\mathbf{Y}$ is a two dimensional vector

$$\mathbf{Y} = \begin{bmatrix} Y_1 \\ Y_2 \end{bmatrix} = \begin{bmatrix} 2X + W_1 \\ X + W_2 \end{bmatrix}.$$

We have

$$\mathbf{C_Y} = \begin{bmatrix} \text{Var}(Y_1) & \text{Cov}(Y_1, Y_2) \\ \text{Cov}(Y_2, Y_1) & \text{Var}(Y_2) \end{bmatrix} = \begin{bmatrix} 22 & 10 \\ 10 & 10 \end{bmatrix},$$

$$\mathbf{C_{XY}} = \begin{bmatrix} \text{Cov}(X, Y_1) & \text{Cov}(X, Y_2) \end{bmatrix} = \begin{bmatrix} 10 & 5 \end{bmatrix}.$$

Therefore,

$$\begin{aligned}
\hat{\mathbf{X}}_L &= \begin{bmatrix} 10 & 5 \end{bmatrix} \begin{bmatrix} 22 & 10 \\ 10 & 10 \end{bmatrix}^{-1} \left(\begin{bmatrix} Y_1 \\ Y_2 \end{bmatrix} - \begin{bmatrix} 0 \\ 0 \end{bmatrix} \right) + 0 \\
&= \begin{bmatrix} \frac{5}{12} & \frac{1}{12} \end{bmatrix} \begin{bmatrix} Y_1 \\ Y_2 \end{bmatrix} \\
&= \frac{5}{12} Y_1 + \frac{1}{12} Y_2,
\end{aligned}$$

which is the same as the result that we obtain using the orthogonality principle in Problem 8.

11. Consider two random variables X and Y with the joint PMF given by the table below.

	$Y=0$	$Y=1$
$X=0$	$\frac{1}{7}$	$\frac{3}{7}$
$X=1$	$\frac{3}{7}$	0

(a) Find the linear MMSE estimator of X given Y, ($\hat{X}_L$).

(b) Find the MMSE estimator of X given Y, ($\hat{X}_M$).

(c) Find the MSE of $\hat{X}_M$.

Solution: Using the table we find out

$$\begin{aligned} P_X(0) &= \frac{1}{7}+\frac{3}{7}=\frac{4}{7}, \\ P_X(1) &= \frac{3}{7}+0=\frac{3}{7}, \\ P_Y(0) &= \frac{1}{7}+\frac{3}{7}=\frac{4}{7}, \\ P_Y(1) &= \frac{3}{7}+0=\frac{3}{7}. \end{aligned}$$

Thus, the marginal distributions of X and Y are both $Bernoulli(\frac{3}{7})$. Therefore, we have

$$\begin{aligned} EX = EY &= \frac{3}{7}, \\ \text{Var}(X)=\text{Var}(Y) &= \frac{3}{7}\cdot\frac{4}{7}=\frac{12}{49}. \end{aligned}$$

(a) To find the linear MMSE estimator of X, given Y, we also need $\text{Cov}(X,Y)$. We have

$$EXY = \sum x_i y_j P_{XY}(x,y) = 0.$$

Therefore,

$$\begin{aligned} \text{Cov}(X,Y) &= EXY - EXEY \\ &= -\frac{9}{49}. \end{aligned}$$

The linear MMSE estimator of X, given Y is

$$\begin{aligned}
\hat{X}_L &= \frac{\text{Cov}(X,Y)}{\text{Var}(Y)}(Y - EY) + EX \\
&= \frac{-9/49}{12/49}\left(Y - \frac{3}{7}\right) + \frac{3}{7} \\
&= -\frac{3}{4}Y + \frac{3}{4}.
\end{aligned}$$

Since Y can only take two values, we can summarize $\hat{X}_L$ in the following table

	$Y = 0$	$Y = 1$
$\hat{X}_L$	$\frac{3}{4}$	0

(b) To find the MMSE estimator of X given Y, we need the conditional PMFs. We have

$$\begin{aligned}
P_{X|Y}(0|0) &= \frac{P_{XY}(0,0)}{P_Y(0)} \\
&= \frac{1}{4}.
\end{aligned}$$

Thus,

$$P_{X|Y}(1|0) = 1 - \frac{1}{4} = \frac{3}{4}.$$

We conclude

$$X|Y = 0 \;\sim\; Bernoulli\left(\frac{3}{4}\right).$$

Similarly, we find

$$\begin{aligned}
P_{X|Y}(0|1) &= 1, \\
P_{X|Y}(1|1) &= 0.
\end{aligned}$$

Thus, given $Y = 1$, we have always $X = 0$. The MMSE estimator of X given Y is

$$\hat{X}_M = E[X|Y].$$

We have

$$E[X|Y=0] = \frac{3}{4},$$
$$E[X|Y=1] = 0.$$

Thus, we can summarize $\hat{X}_M$ in the following table.

Table 9.1: The MMSE estimator of X given Y for Problem 10.

	$Y=0$	$Y=1$
$\hat{X}_M$	$\frac{3}{4}$	0

We notice that, for this problem, the MMSE and the linear MMSE estimators are the same. Here, Y can only take two possible values, and for each value we have a corresponding MMSE estimator. The linear MMSE estimator is just the line passing through the two resulting points.

(c) The MSE of $\hat{X}_M$ can be obtained as

$$\begin{aligned} MSE &= E[\tilde{X}^2] \\ &= EX^2 - E[\hat{X}_M^2] \\ &= \frac{3}{7} - E[\hat{X}_M^2]. \end{aligned}$$

From the table for $\hat{X}_M$, we obtain $E[\hat{X}_M^2] = \frac{4}{7}\left(\frac{3}{4}\right)^2$. Therefore,

$$MSE = \frac{3}{28}.$$

Note that here the MMSE and the linear MMSE estimators are equal, so they have the same MSE. Thus, we can use the formula for the MSE of $\hat{X}_L$ as well

$$
\begin{aligned}
MSE &= (1-\rho(X,Y)^2)\text{Var}(X) \\
&= \left(1-\frac{\text{Cov}(X,Y)^2}{\text{Var}(X)\text{Var}(Y)}\right)\text{Var}(X) \\
&= \left(1-\frac{(-9/49)^2}{12/49\cdot 12/49}\right)\frac{12}{49} \\
&= \frac{3}{28}.
\end{aligned}
$$

13. Suppose that the random variable X is transmitted over a communication channel. Assume that the received signal is given by

$$Y = 2X + W,$$

where $W \sim N(0, \sigma^2)$ is independent of X. Suppose that $X = 1$ with probability p, and $X = -1$ with probability $1-p$. The goal is to decide between $X = -1$ and $X = 1$ by observing the random variable Y. Find the MAP test for this problem.

Solution: Here we have two hypotheses:

H_0: $X = 1$,

H_1: $X = -1$.

Under H_0, $Y = 2 + W$, so $Y|H_0 \sim N(2, \sigma^2)$. Therefore,

$$f_Y(y|H_0) = \frac{1}{\sigma\sqrt{2\pi}}e^{-\frac{(y-2)^2}{2\sigma^2}}.$$

Under H_1, $Y = -2 + W$, so $Y|H_1 \sim N(-2, \sigma^2)$. Therefore,

$$f_Y(y|H_1) = \frac{1}{\sigma\sqrt{2\pi}}e^{-\frac{(y+2)^2}{2\sigma^2}}.$$

Therefore, we choose H_0 if and only if

$$\frac{1}{\sigma\sqrt{2\pi}}e^{-\frac{(y-2)^2}{2\sigma^2}}P(H_0) \geq \frac{1}{\sigma\sqrt{2\pi}}e^{-\frac{(y+2)^2}{2\sigma^2}}P(H_1).$$

We have $P(H_0) = p$, and $P(H_1) = 1 - p$. Therefore, we choose H_0 if and only if

$$\exp\left(\frac{4y}{\sigma^2}\right) \geq \frac{1-p}{p}.$$

Equivalently, we choose H_0 if and only if

$$y \geq \frac{\sigma^2}{4} \ln\left(\frac{1-p}{p}\right).$$

15. A monitoring system is in charge of detecting malfunctioning machinery in a facility. There are two hypotheses to choose from:

    H_0: There is not a malfunction,

    H_1: There is a malfunction.

    The system notifies a maintenance team if it accepts H_1. Suppose that, after processing the data, we obtain $P(H_1|y) = 0.10$. Also, assume that the cost of missing a malfunction is 30 times the cost of a false alarm. Should the system alert a maintenance team (accept H_1)?

    Solution: First, note that

    $$P(H_0|y) = 1 - P(H_1|y) = 0.90.$$

    The posterior risk of accepting H_1 is

    $$P(H_0|y)C_{10} = 0.90C_{10}.$$

    We have $C_{01} = 30C_{10}$, so the posterior risk of accepting H_0 is

    $$\begin{aligned} P(H_1|y)C_{01} &= (0.10)(30C_{10}) \\ &= 3C_{10}. \end{aligned}$$

    Since $P(H_0|y)C_{10} \leq P(H_1|y)C_{01}$, we accept H_1, so an alarm message needs to be sent.

17. When the choice of a prior distribution is subjective, it is often advantageous to choose a prior distribution that will result in a posterior distribution of the same distributional family. When the prior and posterior distributions share the same distributional family, they are called *conjugate distributions*, and the prior is called a *conjugate prior*. Conjugate priors are used out of ease because they always result in a closed form posterior distribution. One example of this is to use a gamma prior for Poisson distributed data.

    Assume our data Y given X is distributed $Y \mid X = x \sim Poisson(\lambda = x)$ and we choose the prior to be $X \sim Gamma(\alpha, \beta)$. Then, the PMF for our data is

    $$P_{Y|X}(y|x) = \frac{e^{-x}x^y}{y!}, \quad \text{for } x > 0, y \in \{0, 12, \dots\},$$

    and the PDF of the prior is given by

    $$f_X(x) = \frac{\beta^\alpha x^{\alpha-1}e^{-\beta x}}{\Gamma(\alpha)}, \quad \text{for } x > 0, \ \ \alpha, \beta > 0.$$

    (a) Show that the posterior distribution is $Gamma(\alpha + y, \beta + 1)$. (*Hint:* Remove all the terms not containing x by putting them into some normalizing constant, c, and noting that $f_{X|Y}(x|y) \propto P_{Y|X}(y|x)f_X(x)$.)

    (b) Write out the PDF for the posterior distribution, $f_{X|Y}(x|y)$.

    (c) Find the mean and the variance of the posterior distribution, $E(X|Y)$ and $Var(X|Y)$.

Solution:

(a)

$$\begin{aligned}
f_{X|Y}(x|y) &\propto P_{Y|X}(y|x)f_X(x) \\
&= \left(\frac{e^{-x}x^y}{y!}\right) \times \left(\frac{\beta^\alpha x^{\alpha-1}e^{-\beta x}}{\Gamma(\alpha)}\right) \\
&= ce^{-x}x^y x^{\alpha-1}e^{-\beta x} && \text{(where c is everything not involving x)} \\
&\propto e^{-x}x^y x^{\alpha-1}e^{-\beta x} && \text{(remove c with proportionality)} \\
&= x^{\alpha+y-1}e^{-x(\beta+1)}.
\end{aligned}$$

This looks like the PDF of a gamma distribution without the normalizing constants. Thus, $f_{X|Y}(x|y) \sim Gamma(\alpha + y, \beta + 1)$.

(b) The posterior PDF is

$$f_{X|Y}(x|y) = \frac{(\beta + 1)^{(\alpha+y)} x^{\alpha+y-1} e^{-(\beta+1)x}}{\Gamma(\alpha + y)}.$$

(c) Since we know the posterior distribution is gamma, $E(X|Y) = \frac{\alpha+y}{\beta+1}$ and $Var(X|Y) = \frac{\alpha+y}{(\beta+1)^2}$.

19. Assume our data Y given X is distributed $Y \mid X = x \sim Geometric(p = x)$ and we chose the prior to be $X \sim Beta(\alpha, \beta)$. Refer to Problem 18 for the PDF and moments of the *Beta* distribution.

(a) Show that the posterior distribution is $Beta(\alpha + 1, \beta + y - 1)$.

(b) Write out the PDF for the posterior distribution, $f_{X|Y}(x|y)$.

(c) Find the mean and the variance of the posterior distribution, $E(X|Y)$ and $Var(X|Y)$.

Solution:

(a)

$$\begin{aligned}
f_{X|Y}(x|y) &\propto P_{Y|X}(y|x) f_X(x) \\
&= \left((1 - x)^{y-1} x\right) \times \left(\frac{\Gamma(\alpha + \beta)}{\Gamma(\alpha)\Gamma(\beta)} x^{\alpha-1} (1 - x)^{\beta-1}\right) \\
&= cx(1 - x)^{y-1} x^{\alpha-1} (1 - x)^{\beta-1} \\
&\propto x(1 - x)^{y-1} x^{\alpha-1} (1 - x)^{\beta-1} \\
&= x^{\alpha} (1 - x)^{\beta+y-2}.
\end{aligned}$$

This looks like the PDF of a beta distribution without the normalizing constants. Thus, $f_{X|Y}(x|y) \sim Beta(\alpha+1, \beta+y-1)$.

(b) The posterior PDF is

$$f_{X|Y}(x|y) = \frac{\Gamma(\alpha+\beta+y)}{\Gamma(\alpha+1)\Gamma(\beta+y-1)} x^{\alpha}(1-x)^{\beta+y-2}.$$

(c) Since the posterior distribution is beta, the mean and variance $E(X|Y) = \frac{\alpha+1}{\alpha+\beta+y}$ and $Var(X|Y) = \frac{(\alpha+1)(\beta+y-1)}{(\alpha+\beta+y)^2(\alpha+\beta+y+1)}$ respectively.

Chapter 10

Introduction to Random Processes

1. Let $\{X_n, n \in \mathbb{Z}\}$ be a discrete-time random process, defined as

$$X_n = 2\cos\left(\frac{\pi n}{8} + \Phi\right),$$

where $\Phi \sim Uniform(0, 2\pi)$.

(a) Find the mean function, $\mu_X(n)$.

(b) Find the correlation function $R_X(m, n)$.

(c) Is X_n a WSS process?

Solution:

(a) We have

$$\begin{aligned}\mu_X(n) &= E[X_n] \\ &= E\left[2\cos\left(\frac{n\pi}{8} + \Phi\right)\right] \\ &= \int_0^{2\pi} 2\cos\left(\frac{\pi n}{8} + \phi\right)\frac{1}{2\pi}d\phi \\ &= 0\end{aligned}$$

(b)

$$\begin{aligned}
R_X(m,n) &= E[4\cos\left(\frac{m\pi}{8}+\Phi\right)\cos\left(\frac{n\pi}{8}+\Phi\right)] \\
&= 2E\left[\cos\left((m-n)\pi/8\right)+\cos\left((m+n)\pi/8+2\Phi\right)\right] \\
&= 2\cos\left(\frac{(m-n)\pi}{8}\right)
\end{aligned}$$

(c) Yes, since $\mu_X(n) = \mu_X$ and $R_X(m,n) = R_X(m-n)$.

3. Let $\{X(n), n \in \mathbb{Z}\}$ be a WSS discrete-time random process with $\mu_X(n) = 1$ and $R_X(m,n) = e^{-(m-n)^2}$. Define the random process $Z(n)$ as

$$Z(n) = X(n) + X(n-1), \qquad \text{for all } n \in \mathbb{Z}.$$

   (a) Find the mean function of $Z(n)$, $\mu_Z(n)$.
   (b) Find the autocorrelation function of $Z(n)$, $R_Z(m,n)$.
   (c) Is $Z(n)$ a WSS random process?

Solution:

(a)

$$\begin{aligned}
\mu_Z(n) &= E[Z(n)] \\
&= E[X(n)] + E[X(n-1)] \\
&= 1+1 \\
&= 2
\end{aligned}$$

(b)

$$\begin{aligned}
R_Z(m,n) &= E[Z(m)\cdot Z(n)] \\
&= E[(X(m)+X(m-1))(X(n)+X(n-1))] \\
&= E[X(m)X(n)] + E[X(m)X(n-1)] + E[X(m-1)X(n)] \\
&+ E[X(m-1)X(n-1)] \\
&= e^{-(m-n)^2} + e^{-(m-n+1)^2} + e^{-(m-1-n)^2} + e^{-(m-n)^2} \\
&= 2e^{-(m-n)^2} + e^{-(m-n+1)^2} + e^{-(m-1-n)^2}
\end{aligned}$$

(c) Yes, since $\mu_Z(n) = \mu_Z$ and $R_Z(m,n) = R_Z(m-n)$.

5. Let $\{X(t), t \in \mathbb{R}\}$ and $\{Y(t), t \in \mathbb{R}\}$ be two independent random processes. Let $Z(t)$ be defined as

$$Z(t) = X(t)Y(t), \qquad \text{for all } t \in \mathbb{R}.$$

Prove the following statements:

(a) $\mu_Z(t) = \mu_X(t)\mu_Y(t)$, for all $t \in \mathbb{R}$.
(b) $R_Z(t_1, t_2) = R_X(t_1, t_2)R_Y(t_1, t_2)$, for all $t \in \mathbb{R}$.
(c) If $X(t)$ and $Y(t)$ are WSS, then they are jointly WSS.
(d) If $X(t)$ and $Y(t)$ are WSS, then $Z(t)$ is also WSS.
(e) If $X(t)$ and $Y(t)$ are WSS, then $X(t)$ and $Z(t)$ are jointly WSS.

Solution:
(a)

$$\begin{aligned}\mu_Z(t) &= E[Z(t)]\\ &= E[X(t)Y(t)]\\ &= E[X(t)]E[Y(t)] \quad \text{(since X and Y are independent)}\\ &= \mu_X(t)\mu_Y(t)\end{aligned}$$

(b)

$$\begin{aligned}R_Z(t_1,t_2) &= E[Z(t_1)\cdot Z(t_2)]\\ &= E[X(t_1)Y(t_1)X(t_2)Y(t_2)]\\ &= E[X(t_1)X(t_2)]E[Y(t_1)Y(t_2)]\\ &= R_X(t_1,t_2)\cdot R_Y(t_1,t_2)\end{aligned}$$

(c)

$$\begin{aligned}R_{XY}(t_1,t_2) &= E[X(t_1)\cdot Y(t_2)]\\ &= E[X(t_1)]E[Y(t_2)]\\ &= \mu_X\cdot\mu_Y \quad \text{(Does not depend on)} \quad t_1, t_2\\ &\text{(You can think of these as a function of} \quad t_1 - t_2)\end{aligned}$$

(d)

$$\begin{aligned}\mu_Z(t) &= \mu_X\mu_Y \quad \text{(By (a) and (b))}\\ R_Z(t_1,t_2) &= R_X(t_1-t_2)R_Y(t_1-t_2)\\ &= R_Z(\tau)\end{aligned}$$

(e) By part (d), $Z(t)$ is also WSS.

$$\begin{aligned}R_{XZ}(t_1,t_2) &= E[X(t_1)\cdot X(t_2)\cdot Y(t_2)]\\ &= E[X(t_1)X(t_2)]E[Y(t_2)]\\ &= R_X(t_1-t_2)\mu_Y\\ &= R_{XZ}(t_1-t_2)\end{aligned}$$

7. Let $X(t)$ be a WSS Gaussian random process with $\mu_X(t) = 1$ and $R_X(\tau) = 1 + 4\text{sinc}(\tau)$.

   (a) Find $P(1 < X(1) < 2)$.

   (b) Find $P(1 < X(1) < 2, X(2) < 3)$.

Solution:
(a) Let $Y = X(1)$, then

$$\begin{aligned}EY &= E[X(1)]\\ &= 1\\ Var(Y) &= R_X(0) - (E[Y])^2\\ &= 5-1=4\\ Y &\sim N(1,4)\\ P(1<Y<2) &= \Phi(\frac{2-1}{2}) - \Phi(\frac{1-1}{2})\\ &= \Phi(\frac{1}{2}) - \Phi(0)\\ &\approx 0.19\end{aligned}$$

(b) Let $Y = X(1)$, $Z = X(2)$. Then Y and Z are jointly Gaussian and $Y \sim N(1,4)$, $Z \sim N(1,4)$.

$$\begin{aligned} Cov(Y,Z) &= E[YZ] - EYEZ \\ &= R_X(-1) - 1 \cdot 1 \\ &= 1 - 1 = 0 \end{aligned}$$

Y and Z are uncorrelated, so Y and Z are independent (jointly Gaussian).

$$\begin{aligned} P(1 < Y < 2, Z < 3) &= P(1 < Y < 2)P(Z < 3) \\ &= [\Phi(\frac{1}{2}) - \Phi(0)][\Phi(\frac{3-1}{2})] \\ &\approx 0.16 \end{aligned}$$

9. Let $\{X(t), t \in \mathbb{R}\}$ be a continuous-time random process, defined as

$$X(t) = \sum_{k=0}^{n} A_k t^k,$$

where A_0, A_1, $\cdots$, A_n are i.i.d. $N(0,1)$ random variables and n is a fixed positive integer.

(a) Find the mean function $\mu_X(t)$.

(b) Find the correlation function $R_X(t_1, t_2)$.

(c) Is $X(t)$ a WSS process?

(d) Find $P(X(1) < 1)$. Assume $n = 10$.

(e) Is $X(t)$ a Gaussian process?

Solution:

(a)

$$\begin{aligned}\mu_X(t) &= E\left[\sum_{k=0}^{n} A_k t^k\right]\\ &= \sum_{k=0}^{n} E[A_k]t^k\\ &= 0\end{aligned}$$

(b)

$$\begin{aligned}R_X(t_1,t_2) &= E[X(t_1)X(t_2)]\\ &= E[\sum_{k=0}^{n} A_k t_1^k \sum_{l=0}^{n} A_l t_2^l]\\ &= \sum_{k=0}^{n}\sum_{l=0}^{n} E[A_k A_l] t_1^k t_2^l\\ &= \sum_{k=0}^{n} E[A_k^2] t_1^k t_2^k\\ &= \sum_{k=0}^{n} (t_1 t_2)^k\end{aligned}$$

(c) No, since $R_X(t_1,t_2) \neq R_X(t_1 - t_2)$.

(d)

$$\begin{aligned}n &= 10\\ X(t) &= \sum_{k=0}^{10} A_k t^k\\ X(1) &= \sum_{k=0}^{10} A_k \quad A_k \sim N(0,1)(i.i.d)\\ X(1) &\sim N(0,10)\\ P(X(1)<1) &= \Phi\left(\frac{1-0}{\sqrt{10}}\right)\\ &= \Phi\left(\frac{1}{\sqrt{10}}\right)\\ &= 0.624\end{aligned}$$

(e) Yes, since any linear combination of

$$X(t_1), X(t_2), X(t_3), \cdots, X(t_l)$$

can be written as a linear combination of

$$A_0, A_1, A_2, \cdots, A_n$$

Since $A_0, A_1, \cdots, A_n$ are jointly normal, we conclude that $X(t_1)$, $\cdots$, $X(t_l)$ are jointly normal.

11. (Time Averages) Let $\{X(t), t \in \mathbb{R}\}$ be a continuous-time random process. The time average mean of $X(t)$ is defined as [1]

$$\langle X(t) \rangle = \lim_{T \to \infty} \left[\frac{1}{2T} \int_{-T}^{T} X(t) dt \right].$$

Consider the random process $\big\{X(t), t \in \mathbb{R}\big\}$ defined as

$$X(t) = \cos(t + U),$$

where $U \sim Uniform(0, 2\pi)$. Find $\langle X(t) \rangle$.

Solution:

Let $U = u$. So $X(t) = \cos(t + u)$. Note that

$$\int_{-T}^{T} \cos(t + u) dt = \sin(T + u) - \sin(-T + u)$$

$$\left| \int_{-T}^{T} \cos(t + u) dt \right| \leq 2$$

$$\left| \frac{1}{2T} \int_{-T}^{T} \cos(t + u) dt \right| \leq \frac{1}{T}$$

$$\lim_{T \to \infty} [\frac{1}{2T} \int_{-T}^{T} X(t) dt] = 0$$

[1] Assuming that the limit exists in mean-square sense.

13. Let $\{X(t), t \in \mathbb{R}\}$ be a WSS random process. Show that for any $\alpha > 0$, we have

$$P\big(|X(t+\tau) - X(t)| > \alpha\big) \leq \frac{2R_X(0) - 2R_X(\tau)}{\alpha^2}.$$

Solution: Let $Y = X(t+\tau) - X(t)$. Then,

$$\begin{aligned}
EY &= E[X(t+\tau) - X(t)] = 0 \\
\mathrm{Var}(Y) &= E[Y^2] \\
&= E[X^2(t+\tau) + X^2(t) - 2X(t+\tau)X(t)] \\
&= R_X(0) + R_X(0) - 2R_X(\tau) \\
&= 2R_X(0) - 2R_X(\tau) \\
&= P(|Y - 0| > \alpha) \leq \frac{\mathrm{Var}(Y)}{\alpha^2} \quad \text{(Chebyshev's Inequality)} \\
&= \frac{2R_X(0) - 2R_X(\tau)}{\alpha^2}
\end{aligned}$$

15. Let $X(t)$ be a real-valued WSS random process with autocorrelation function $R_X(\tau)$. Show that the Power Spectral Density (PSD) of $X(t)$ is given by

$$S_X(f) = \int_{-\infty}^{\infty} R_X(\tau) \cos(2\pi f \tau) \; d\tau.$$

Solution:

$$\begin{aligned}
S_X(f) &= \mathcal{F}\{R_X(\tau)\} \\
&= \int_{-\infty}^{\infty} R_X(\tau) e^{-2j\pi f \tau} \; d\tau \\
&= \int_{-\infty}^{\infty} R_X(\tau)(\cos 2\pi f_c \tau - j \sin 2\pi f_c \tau) d\tau \\
&= \int_{-\infty}^{\infty} R_X(\tau) \cos(2\pi f_c \tau) d\tau - j \int_{-\infty}^{\infty} R_X(\tau) \sin(2\pi f_c \tau) d\tau \\
&= \int_{-\infty}^{\infty} R_X(\tau) \cos(2\pi f_c \tau) d\tau
\end{aligned}$$

The integral $\int_{-\infty}^{\infty} R_X(\tau)\sin(2\pi f_c\tau)d\tau$ is equal to zero, since $R_X(\tau)$ is an even function and $\sin(2\pi f_c)$ is an odd function. Therefore, $R_X(\tau)\sin(2\pi f\tau)d\tau$ is an odd function.

17. Let $X(t)$ be a WSS process with autocorrelation function

$$R_X(\tau) = \frac{1}{1+\pi^2\tau^2}.$$

Assume that $X(t)$ is input to a low-pass filter with frequency response

$$H(f) = \begin{cases} 3 & |f| < 2 \\ 0 & \text{otherwise} \end{cases}$$

Let $Y(t)$ be the output.

(a) Find $S_X(f)$.

(b) Find $S_{XY}(f)$.

(c) Find $S_Y(f)$.

(d) Find $E[Y(t)^2]$.

Solution:

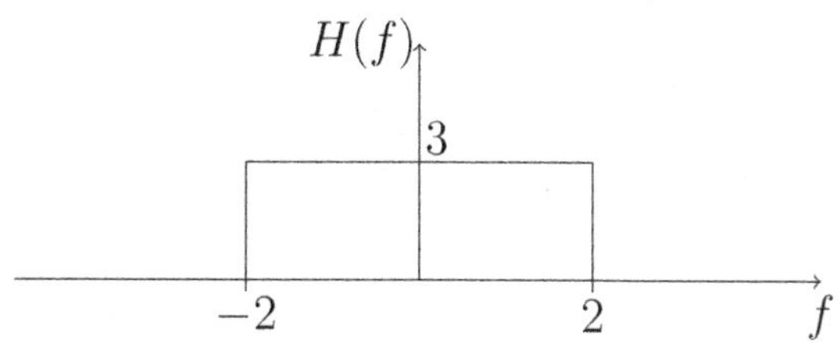

Figure 10.1: A lowpass filter

$$R_X(\tau) = \frac{1}{1+\pi^2\tau^2}.$$

(a)

$$S_X(f) = \mathcal{F}\{\frac{1}{1+\pi^2\tau^2}\}$$
$$= e^{-2|f|} \quad \text{for all} \quad f \in \mathbb{R}$$

(b)

$$S_{XY}(f) = S_X(f)H^*(f) = \begin{cases} 3e^{-2|f|} & |f| < 2 \\ 0 & \text{else} \end{cases}$$

(c)

$$S_Y(f) = S_X(f)|H(f)|^2 = \begin{cases} 9e^{-2|f|} & |f| < 2 \\ 0 & \text{else} \end{cases}$$

(d)

$$\begin{aligned} E[Y(t)^2] &= \int_{-\infty}^{\infty} S_Y(f) \; df \\ &= \int_{-2}^{2} 9e^{-2|f|} df \\ &= 2\int_{0}^{2} 9e^{-2f} df \\ &= 9\left(1 - e^{-4}\right) \end{aligned}$$

19. Let $X(t)$ be a zero-mean WSS Gaussian random process with $R_X(\tau) = e^{-\pi\tau^2}$. Suppose that $X(t)$ is input to an LTI system with transfer function

$$|H(f)| = e^{-\frac{3}{2}\pi f^2}.$$

Let $Y(t)$ be the output.

(a) Find μ_Y.

(b) Find $R_Y(\tau)$ and $\mathrm{Var}(Y(t))$.

(c) Find $E[Y(3)|Y(1) = -1]$.

(d) Find $\mathrm{Var}(Y(3)|Y(1) = -1)$.

(e) Find $P(Y(3) < 0|Y(1) = -1)$.

Solution:

(a)

$$\begin{aligned}\mu_Y &= \mu_X H(0)\\ &= 0\end{aligned}$$

(b)

$$\begin{aligned}S_Y(f) &= S_X(f)|H(f)|^2\\ &= e^{-\pi f^2}|H(f)|^2\\ &= e^{-4\pi f^2}\\ R_Y(\tau) &= \mathcal{F}^{-1}\{S_Y(f)\}\\ &= \mathcal{F}^{-1}\{e^{-\pi(2f)^2}\}\\ &= \frac{1}{2}e^{-\pi\left(\frac{\tau}{2}\right)^2}\\ \mathrm{Var}(Y(t)) &= E[Y(t)^2]\\ &= R_Y(0)\\ &= \frac{1}{2}\end{aligned}$$

(c) $Y(3)$ and $Y(1)$ are zero-mean jointly normal random variables.

$$\begin{aligned}
\text{Cov}(Y(3), Y(1)) &= E[Y(3)Y(1)] \\
&= R_Y(2) \\
&= \frac{1}{2}e^{-\pi}
\end{aligned}$$

$$\begin{aligned}
E[Y(3)|Y(1) = -1] &= E[Y(3)] + \frac{cov(Y(3), Y(1))}{\text{Var}(Y(1))}(-1-0) \\
&= 0 + \frac{\frac{1}{2}e^{-\pi}}{\frac{1}{2}}(-1) \\
&= -e^{-\pi}
\end{aligned}$$

(d)

$$\begin{aligned}
\text{Var}(Y(3)|Y(1) = -1) &= (1-\rho^2)\text{Var}(Y(3)) \\
\rho &= \frac{\text{Cov}(Y(3), Y(1))}{\sqrt{\text{Var}(Y(3))\text{Var}(Y(1))}} \\
&= \frac{\frac{1}{2}e^{-\pi}}{\frac{1}{2}} \\
&= e^{-\pi} \\
\text{Var}(Y(3)|Y(1) = -1) &= (1-e^{-2\pi})\frac{1}{2}
\end{aligned}$$

(e) $Y(3)|Y(1) = -1 \sim N\left(-e^{-\pi}, \frac{1-e^{-2\pi}}{2}\right)$. Thus,

$$\begin{aligned}
P(Y(3) < 0|Y(1) = -1) &= \Phi\left(\frac{0 + e^{-\pi}}{\sqrt{\frac{1-e^{-2\pi}}{2}}}\right) \\
&= 0.5244
\end{aligned}$$

Chapter 11

Some Important Random Processes

1. The number of orders arriving at a service facility can be modeled by a Poisson process with intensity $\lambda = 10$ orders per hour.

   (a) Find the probability that there are no orders between 10:30 and 11:00.

   (b) Find the probability that there are 3 orders between 10:30 and 11:00 and 7 orders between 11:30 and 12:00.

Solution:

(a) Let $X = N(11) - N(10.5)$, then $X \sim Poisson(10 \cdot \frac{1}{2})$, thus $P(X = 0) = e^{-5}$.

(b) Let

$$
\begin{aligned}
X_1 &= N(11) - N(10.5) \\
X_2 &= N(12) - N(11.5)
\end{aligned}
$$

Then X_1 and X_2 are two independent $Poisson(5)$ random variables. So

$$
\begin{aligned}
P(X_1 = 3, X_2 = 7) &= P(X_1 = 3)P(X_2 = 7) \\
&= \frac{e^{-5}5^3}{3!} \cdot \frac{e^{-5}5^7}{7!}
\end{aligned}
$$

3. Let $X \sim Poisson(\mu_1)$ and $Y \sim Poisson(\mu_2)$ be two independent random variables. Define $Z = X + Y$.
Show that

$$X|Z = n \sim Binomial\left(n, \frac{\mu_1}{\mu_1 + \mu_2}\right).$$

Solution: First note that

$$Z = X + Y \sim Poisson(\mu_1 + \mu_2).$$

We can write

$$\begin{aligned}
P(X = k|Z = n) &= \frac{P(X = k, Z = n)}{P(Z = n)} \\
&= \frac{P(X = k, Y = n - k)}{P(Z = n)} \\
&= \frac{P(X = k)P(Y = n - k)}{P(Z = n)} \\
&= \frac{\frac{e^{-\mu_1}(\mu_1)^k}{k!} \frac{e^{-\mu_2}(\mu_2)^{n-k}}{(n-k)!}}{\frac{e^{-(\mu_1+\mu_2)}(\mu_1+\mu_2)^n}{n!}} \\
&= \binom{n}{k} \left(\frac{\mu_1}{\mu_1 + \mu_2}\right)^k \left(1 - \frac{\mu_1}{\mu_1 + \mu_2}\right)^{(n-k)} \\
X|Z = n &\sim Binomial\left(n, \frac{\mu_1}{\mu_1 + \mu_2}\right)
\end{aligned}$$

5. Let $N_1(t)$ and $N_2(t)$ be two independent Poisson processes with rate λ_1 and λ_2 respectively. Let $N(t) = N_1(t) + N_2(t)$ be the merged process. Show that given $N(t) = n$, $N_1(t) \sim Binomial\left(n, \frac{\lambda_1}{\lambda_1+\lambda_2}\right)$.
Note: We can interpret this result as follows: Any arrival in the merged process belongs to $N_1(t)$ with probability $\frac{\lambda_1}{\lambda_1+\lambda_2}$ and belongs to $N_2(t)$ with

probability $\frac{\lambda_2}{\lambda_1+\lambda_2}$ independent of other arrivals.

Solution: This is the direct result of problem 3. Here we have

$$\begin{aligned}
X &= N_1(t) \\
Y &= N_2(t) \\
X &\sim Poisson(\eta_1 = \lambda_1 t) \\
Y &\sim Poisson(\eta_2 = \lambda_2 t) \\
Z &\sim Poisson\left(\eta = \eta_1 + \eta_2\right) \\
\text{Thus,} \quad X|Z = n &\sim Binomial(n, \frac{\eta_1}{\eta_1 + \eta_2}) \\
&= Binomial\left(n, \frac{\lambda_1}{\lambda_1 + \lambda_2}\right)
\end{aligned}$$

7. Let $\{N(t), t \in [0, \infty)\}$ be a Poisson Process with rate λ. Let T_1, T_2, $\cdots$ be the arrival times for this process. Show that

$$f_{T_1,T_2,\ldots,T_n}(t_1, t_2, \cdots, t_n) = \lambda^n e^{-\lambda t_n}, \qquad \text{for } 0 < t_1 < t_2 < \cdots < t_n.$$

Hint: One way to show the above result is to show that for sufficiently small Δ_i, we have

$$P\Big(t_1 \leq T_1 < t_1 + \Delta_1, t_2 \leq T_2 < t_2 + \Delta_2, ..., t_n \leq T_n < t_n + \Delta_n) \approx$$
$$\lambda^n e^{-\lambda t_n} \Delta_1 \Delta_2 \cdots \Delta_n, \qquad \text{for } 0 < t_1 < t_2 < \cdots < t_n.$$

Solution:

$$P(t_i \leq T_i < t_i + \Delta_i) \quad \text{for} \quad (i = 1, 2, \cdots, n)$$

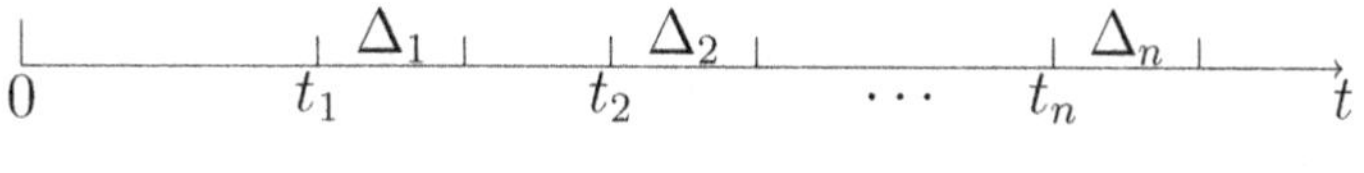

Figure 11.1:

$$
\begin{aligned}
&P\left(t_1 \leq T_1 < t_1 + \Delta_1, \cdots, t_n \leq T_n < t_n + \Delta_n\right) \\
&= P[\text{one arrival in} \quad [t_1, t_1 + \Delta), \cdots, \quad \text{one arrival in} \quad [t_n, t_n + \Delta)] \\
&\times P[\text{no arrivals in} \quad [0, t_1), \quad \text{no arrivals in} \quad [t_1 + \Delta, t_2), \cdots] \\
&= \left(\lambda\Delta_1 e^{-\lambda\Delta_1}\right) \cdots \left(\lambda\Delta_n e^{-\lambda\Delta_n}\right) \cdot e^{-\lambda(t - \Delta_1 - \Delta_2 - \cdots - \Delta_n)} \\
&= \lambda^n e^{-\lambda(\Delta_1 + \cdots + \Delta_n)} \cdot e^{-\lambda(t_n - (\Delta_1 + \cdots + \Delta_n))} \left(\Delta_1 \cdots \Delta_n\right) \\
&= \lambda^n e^{-\lambda t_n} \cdot \Delta_1 \cdots \Delta_n.
\end{aligned}
$$

Therefore,

$$
\begin{aligned}
P\left(t_1 \leq T_1 < t_1 + \Delta_1, \cdots, t_n \leq T_n < t_n + \Delta_n\right) &\approx f_{T_1, \cdots, T_n}\left(t_1, \cdots, t_n\right) \cdot \Delta_1 \cdots \Delta_n \\
&= \lambda^n e^{-\lambda t_n} \cdot \Delta_1 \cdots \Delta_n.
\end{aligned}
$$

We conclude

$$
f_{T_1, \cdots, T_n}\left(t_1, \cdots, t_n\right) = \lambda^n e^{-\lambda t_n} \quad \text{for} \quad 0 < t_1 < t_2 < \cdots < t_n
$$

9. Let $\{N(t), t \in [0, \infty)\}$ be a Poisson Process with rate λ. Let T_1, T_2, $\cdots$ be the arrival times for this process. Find

$$
E[T_1 + T_2 + \cdots + T_{10} | N(4) = 10].
$$

Hint: Use the result of Problem 8.

Solution: By Problem 8, we can say:

Given $N(4) = 10$, then $T_1 + \cdots + T_{10}$ has the same distribution as $U = U_1 + U_2 + \cdots + U_{10}$ where $U_i \sim Uniform(0, 4)$ and U_i's are independent. Thus:

$$
\begin{aligned}
E\left[T_1 + \cdots + T_{10} | N(4) = 10\right] &= E\left[U_1 + \cdots + U_{10}\right] \\
&= 10 E\left[U_i\right] \\
&= 20
\end{aligned}
$$

11. In Problem 10, find the probability that Team B scores the first goal. That is, find the probability that at least one goal is scored in the game and the first goal is scored by Team B.

Solution:

Given that the first goal is scored at $t \leq 90$, then the goal is scored by team B with probability $\frac{\lambda_2}{\lambda_1+\lambda_2} = \frac{3}{5}$ (see Problem 5). The probability of scoring at least one goal is

$$P[N(90) > 0] = 1 - e^{-4.5}$$

Thus the desired probability is

$$\left(1 - e^{-4.5}\right)\frac{3}{5}.$$

13. Consider the Markov chain with three states $S = \{1, 2, 3\}$, that has the state transition diagram as shown in Figure 11.31.

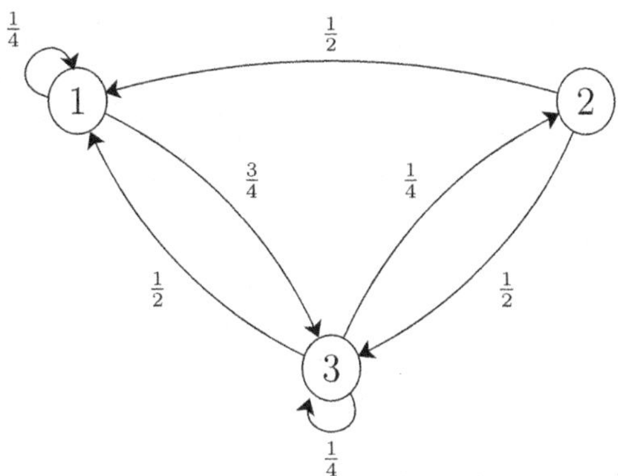

Figure 11.2: A state transition diagram.

Suppose $P(X_1 = 1) = \frac{1}{2}$ and $P(X_1 = 2) = \frac{1}{4}$.

(a) Find the state transition matrix for this chain.

(b) Find $P(X_1 = 3, X_2 = 2, X_3 = 1)$.

(c) Find $P(X_1 = 3, X_3 = 1)$.

Solution:

(a) The state transition matrix is given by

$$P = \begin{bmatrix} \frac{1}{4} & 0 & \frac{3}{4} \\ \frac{1}{2} & 0 & \frac{1}{2} \\ \frac{1}{2} & \frac{1}{4} & \frac{1}{4} \end{bmatrix}.$$

(b) First, we obtain

$$\begin{aligned} P(X_1 = 3) &= 1 - P(X_1 = 1) - P(X_1 = 2) \\ &= 1 - \frac{1}{2} - \frac{1}{4} \\ &= \frac{1}{4}. \end{aligned}$$

We can now write

$$\begin{aligned} P(X_1 = 3, X_2 = 2, X_3 = 1) &= P(X_1 = 3) \cdot p_{32} \cdot p_{21} \\ &= \frac{1}{4} \cdot \frac{1}{4} \cdot \frac{1}{2} \\ &= \frac{1}{32}. \end{aligned}$$

(c) We can write

$$\begin{aligned} P(X_1 = 3, X_3 = 1) &= \sum_{k=1}^{3} P(X_1 = 3, X_2 = k, X_3 = 1) \\ &= \sum_{k=1}^{3} P(X_1 = 3) \cdot p_{3k} \cdot p_{k1} \\ &= P(X_1 = 3)\big[p_{31} \cdot p_{11} + p_{32} \cdot p_{21} + p_{33} \cdot p_{31}\big] \\ &= \frac{1}{4}\big[\frac{1}{2} \cdot \frac{1}{4} + \frac{1}{4} \cdot \frac{1}{2} + \frac{1}{4} \cdot \frac{1}{2}\big] \\ &= \frac{3}{32}. \end{aligned}$$

15. Let X_n be a discrete-time Markov chain. Remember that, by definition, $p_{ii}^{(n)} = P(X_n = i|X_0 = i)$. Show that state i is recurrent if and only if

$$\sum_{n=1}^{\infty} p_{ii}^{(n)} = \infty.$$

Solution: Let V be the total number of visits to state i. Define the random variables Y_n's as follows:

$$Y_n = \begin{cases} 1 & \text{if } X_n = i \\ 0 & \text{otherwise} \end{cases}$$

Then, we have

$$V = \sum_{n=0}^{\infty} Y_n.$$

Therefore,

$$\begin{aligned} E[V|X_0 = i] &= \sum_{n=0}^{\infty} E[Y_n = i|X_0 = i] \\ &= \sum_{n=0}^{\infty} P(X_n = i|X_0 = i) \\ &= 1 + \sum_{n=1}^{\infty} p_{ii}^{(n)}. \end{aligned}$$

Now, as we have seen in the text, i is a recurrent state if and only if $E[V|X_0 = i] = \infty$. We conclude that state i is recurrent if and only if

$$\sum_{n=1}^{\infty} p_{ii}^{(n)} = \infty.$$

17. Consider the Markov chain of Problem 16. Again assume $X_0 = 4$. We would like to find the expected time (number of steps) until the chain gets absorbed in R_1 or R_2. More specifically, let T be the absorption time, i.e., the first time the chain visits a state in R_1 or R_2. We would like to find $E[T|X_0 = 4]$.

Solution: Here, we follow our standard procedure for finding mean hitting times. Consider Figure 11.3.

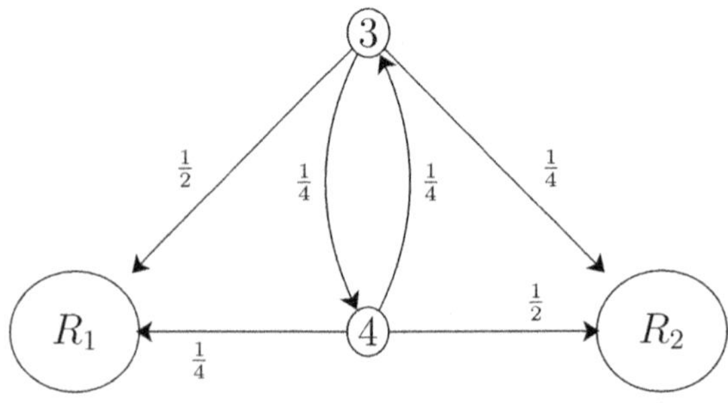

Figure 11.3: The state transition diagram in which we have replaced each recurrent class with one absorbing state

Let T be the first time the chain visits R_1 or R_2. For all $i \in S$, define

$$t_i = E[T|X_0 = i].$$

By the above definition, we have $t_{R_1} = t_{R_2} = 0$. To find t_3 and t_4, we can use the following equations

$$t_i = 1 + \sum_k t_k p_{ik}, \qquad \text{for } i = 3, 4.$$

Specifically, we obtain

$$\begin{aligned} t_3 &= 1 + \frac{1}{2}t_{R_1} + \frac{1}{4}t_4 + \frac{1}{4}t_{R_2} \\ &= 1 + \frac{1}{4}t_4, \end{aligned}$$

$$\begin{aligned} t_4 &= 1 + \frac{1}{4}t_{R_1} + \frac{1}{4}t_3 + \frac{1}{2}t_{R_2} \\ &= 1 + \frac{1}{4}t_3. \end{aligned}$$

Solving the above equations, we obtain

$$t_3 = \frac{4}{3}, \qquad t_4 = \frac{4}{3}.$$

Therefore, if $X_0 = 4$, it will take on average $t_4 = \frac{4}{3}$ steps until the chain gets absorbed in R_1 or R_2.

19. Consider the Markov chain shown in Figure 11.34.

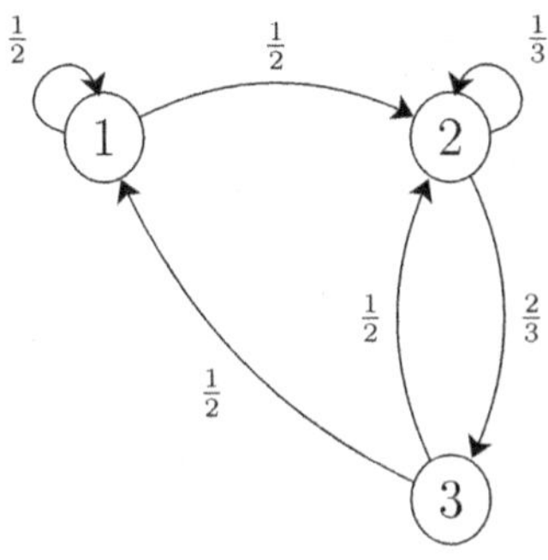

Figure 11.4: A state transition diagram.

(a) Is this chain irreducible?

(b) Is this chain aperiodic?

(c) Find the stationary distribution for this chain.

(d) Is the stationary distribution a limiting distribution for the chain?

Solution:

(a) The chain is irreducible since we can go from any state to any other state in a finite number of steps.

(b) The chain is aperiodic since there is a self-transition, e.g., $p_{11} > 0$.

(c) To find the stationary distribution, we need to solve

$$\pi_1 = \frac{1}{2}\pi_1 + \frac{1}{2}\pi_3,$$
$$\pi_2 = \frac{1}{2}\pi_1 + \frac{1}{3}\pi_2 + \frac{1}{2}\pi_3,$$
$$\pi_3 = \frac{2}{3}\pi_2,$$
$$\pi_1 + \pi_2 + \pi_3 = 1.$$

We find

$$\pi_1 = \frac{2}{7}, \quad \pi_2 = \frac{3}{7}, \quad \pi_3 = \frac{2}{7}.$$

(d) The above stationary distribution is a limiting distribution for the chain because the chain is both irreducible and aperiodic.

21. Consider the Markov chain shown in Figure 11.36. Assume that $0 < p < q$. Does this chain have a limiting distribution? For all $i, j \in \{0, 1, 2, \cdots\}$, find

$$\lim_{n\to\infty} P(X_n = j | X_0 = i).$$

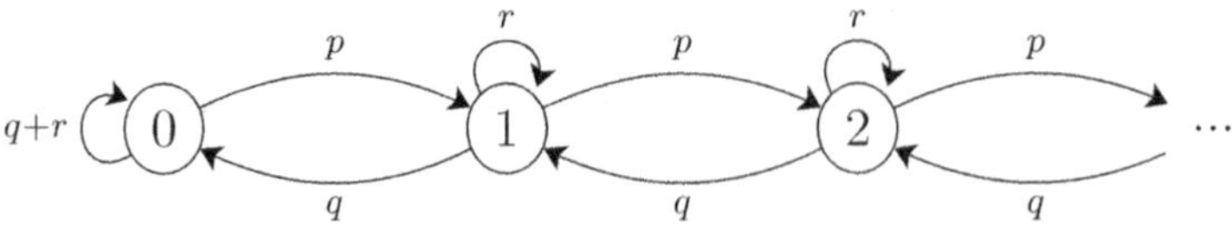

Figure 11.5: A state transition diagram.

Solution: This chain is irreducible since all states communicate with each other. It is also aperiodic since it includes self-transitions. Note that we have $p + q + r = 1$. Let's write the equations for a stationary distribution. For state 0, we can write

$$\pi_0 = (q + r)\pi_0 + q\pi_1,$$

which results in

$$\pi_1 = \frac{p}{q}\pi_0.$$

For state 1, we can write

$$\begin{aligned}\pi_1 &= r\pi_1 + p\pi_0 + q\pi_2 \\ &= r\pi_1 + q\pi_1 + q\pi_2,\end{aligned}$$

which results in

$$\pi_2 = \frac{p}{q}\pi_1.$$

Similarly, for any $j \in \{1, 2, \cdots\}$, we obtain

$$\pi_j = \alpha\pi_{j-1},$$

where $\alpha = \frac{p}{q}$. Note that since $0 < p < q$, we conclude that $0 < \alpha < 1$. We conclude

$$\pi_j = \alpha^j \pi_0, \qquad \text{for } j = 1, 2, \cdots.$$

Finally, we must have

$$\begin{aligned}1 &= \sum_{j=0}^{\infty} \pi_j \\ &= \sum_{j=0}^{\infty} \alpha^j \pi_0, & \text{(where } 0 < \alpha < 1\text{)} \\ &= \frac{1}{1-\alpha}\pi_0 & \text{(geometric series).}\end{aligned}$$

Thus, $\pi_0 = 1 - \alpha$. Therefore, the stationary distribution is given by

$$\pi_j = (1 - \alpha)\alpha^j, \qquad \text{for } j = 0, 1, 2, \cdots.$$

Since this chain is both irreducible and aperiodic and we have found a stationary distribution, we conclude that all states are positive recurrent and $\pi = [\pi_0, \pi_1, \cdots]$ is the limiting distribution.

23. (Gambler's Ruin Problem) Two gamblers, call them Gambler A and Gambler B, play repeatedly. In each round, A wins 1 dollar with probability p or loses 1 dollar with probability $q = 1-p$ (thus, equivalently, in each round B wins 1 dollar with probability $q = 1-p$ and loses 1 dollar with probability p). We assume different rounds are independent. Suppose that, initially, A has i dollars and B has $N-i$ dollars. The game ends when one of the gamblers runs out of money (in which case the other gambler will have N dollars). Our goal is to find p_i, the probability that A wins the game given that he has initially i dollars.

    (a) Define a Markov chain as follows: The chain is in state i if the Gambler A has i dollars. Here, the state space is $S = \{0, 1, \cdots, N\}$. Draw the state transition diagram of this chain.

    (b) Let a_i be the probability of absorption to state N (the probability that A wins) given that $X_0 = i$. Show that

    $$\begin{aligned} a_0 &= 0, \\ a_N &= 1, \\ a_{i+1} - a_i &= \frac{q}{p}(a_i - a_{i-1}), \qquad \text{for } i = 1, 2, \cdots, N-1. \end{aligned}$$

    (c) Show that

    $$a_i = \left[1 + \frac{q}{p} + \left(\frac{q}{p}\right)^2 + \cdots + \left(\frac{q}{p}\right)^{i-1}\right] a_1, \text{ for } i = 1, 2, \cdots, N.$$

    (d) Find a_i for any $i \in \{0, 1, 2, \cdots, N\}$. Consider two cases: $p = \frac{1}{2}$ and $p \neq \frac{1}{2}$.

Solution:

(a) The state transition diagram of the chain is shown in Figure 11.6.

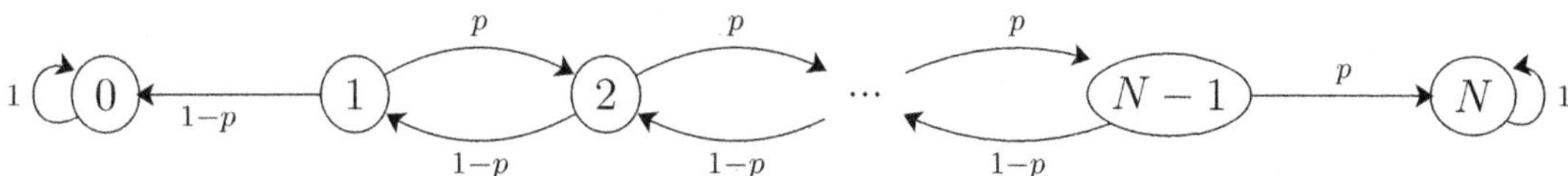

Figure 11.6: The state transition diagram for the gambler's ruin problem.

(b) Applying the law of total probability, we conclude that

$$a_i = pa_{i+1} + (1-p)a_{i-1}, \qquad \text{for } i = 1, 2, \cdots, N-1.$$

Since states 0 and N are absorbing, we conclude that

$$a_0 = 0,$$
$$a_N = 1.$$

From the above, we conclude

$$a_{i+1} = \frac{a_i}{p} - \frac{1-p}{p} a_{i-1}, \qquad \text{for } i = 1, 2, \cdots, N-1.$$

Thus,

$$a_{i+1} - a_i = \frac{q}{p}(a_i - a_{i-1}), \qquad \text{for } i = 1, 2, \cdots, N-1.$$

(c) For $i = 1$, we obtain

$$a_2 - a_1 = \frac{q}{p}(a_1 - a_0) = \frac{q}{p} a_1.$$

Thus,

$$a_2 = \left[1 + \frac{q}{p}\right] a_1.$$

Similarly,

$$a_3 - a_2 = \frac{q}{p}(a_2 - a_1) = \left(\frac{q}{p}\right)^2 a_1.$$

Thus,

$$\begin{aligned} a_3 &= a_2 + \left(\frac{q}{p}\right)^2 a_1 \\ &= \left[1 + \frac{q}{p}\right] a_1 + \left(\frac{q}{p}\right)^2 a_1 \\ &= \left[1 + \frac{q}{p} + \left(\frac{q}{p}\right)^2\right] a_1. \end{aligned}$$

And, so on. In general, we obtain

$$a_i = \left[1 + \frac{q}{p} + \left(\frac{q}{p}\right)^2 + \cdots + \left(\frac{q}{p}\right)^{i-1}\right] a_1, \text{ for } i = 1, 2, \cdots, N.$$

(d) Using the above, we obtain

$$a_N = \left[1 + \frac{q}{p} + \left(\frac{q}{p}\right)^2 + \cdots + \left(\frac{q}{p}\right)^{N-1}\right] a_1.$$

Since, $a_N = 1$, we conclude

$$a_1 = \frac{1}{\left[1 + \frac{q}{p} + \left(\frac{q}{p}\right)^2 + \cdots + \left(\frac{q}{p}\right)^{N-1}\right]}.$$

We thus have

$$a_i = \left[1 + \frac{q}{p} + \left(\frac{q}{p}\right)^2 + \cdots + \left(\frac{q}{p}\right)^{i-1}\right] a_1, \text{ for } i = 1, 2, \cdots, N.$$

We can obtain a_i for any i. Specifically, we obtain

$$a_i = \begin{cases} \frac{1-\left(\frac{q}{p}\right)^i}{1-\left(\frac{q}{p}\right)^N} \text{ if } p \neq \frac{1}{2} \\ \\ \frac{i}{N} \text{ if } p = \frac{1}{2} \end{cases}$$

25. The Poisson process is a continuous-time Markov chain. Specifically, let $N(t)$ be a Poisson process with rate λ.

    (a) Draw the state transition diagram of the corresponding jump chain.

    (b) What are the rates λ_i for this chain?

Solution: Here, the process starts at state 0 ($N(0) = 0$). It stays at state 0 for some time and then moves to state 1. In general, the process goes from state i to state $i + 1$. Thus, the jump chain can be shown by Figure 11.7.

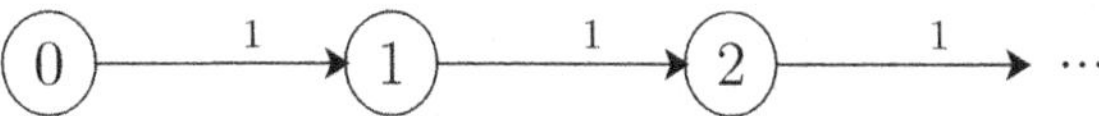

Figure 11.7: The jump chain for the Poisson process.

Remember that the interarrival times in the Poisson process have *Exponential*(λ) distribution. Thus, the time that the chain spends at each state has *Exponential*(λ) distribution. We conclude that

$$\lambda_i = \lambda.$$

27. Consider a continuous-time Markov chain $X(t)$ that has the jump chain shown in Figure 11.8. Assume $\lambda_1 = 1$, $\lambda_2 = 2$, and $\lambda_3 = 4$.

    (a) Find the generator matrix for this chain.

    (b) Find the limiting distribution for $X(t)$ by solving $\pi G = 0$.

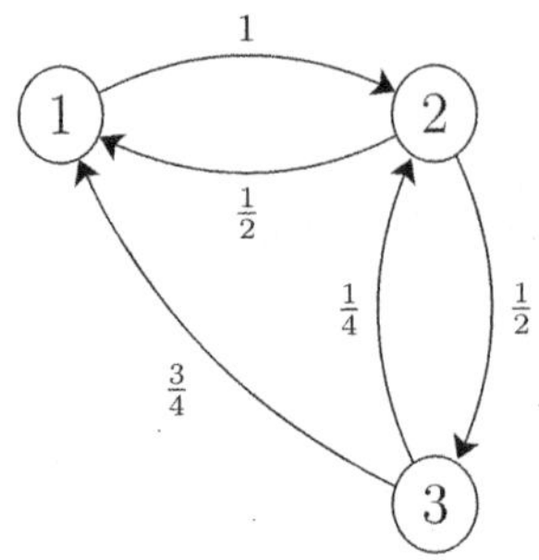

Figure 11.8: The jump chain for the Markov chain of Problem 27.

Solution: The jump chain is irreducible and the transition matrix of the jump chain is given by

$$P = \begin{bmatrix} 0 & 1 & 0 \\ \frac{1}{2} & 0 & \frac{1}{2} \\ \frac{3}{4} & \frac{1}{4} & 0 \end{bmatrix}.$$

The generator matrix can be obtained using

$$g_{ij} = \begin{cases} \lambda_i p_{ij} & \text{if } i \neq j \\ \\ -\lambda_i & \text{if } i = j \end{cases}$$

We obtain

$$G = \begin{bmatrix} -1 & 1 & 0 \\ \\ 1 & -2 & 1 \\ \\ 3 & 1 & -4 \end{bmatrix}.$$

Solving

$$\pi G = 0, \qquad \text{and} \qquad \pi_1 + \pi_2 + \pi_3 = 1$$

we obtain $\pi = \frac{1}{12}[7, 4, 1]$.

29. Let $W(t)$ be the standard Brownian motion.

(a) Find $P(-1 < W(1) < 1)$.

(b) Find $P(1 < W(2) + W(3) < 2)$.

(c) Find $P(W(1) > 2|W(2) = 1)$.

Solution:

(a) Note that $W(1) \sim N(0, 1)$, thus

$$\begin{aligned} P(-1 < W(1) < 1) &= \Phi\left(\frac{1-0}{1}\right) - \Phi\left(\frac{-1-0}{1}\right) \\ &= \Phi(1) - \Phi(-1) \\ &\approx 0.68 \end{aligned}$$

(b) Let $X = W(2) + W(3)$. Then, X is normal with $EX = 0$ and

$$\begin{aligned} \text{Var}(X) &= \text{Var}\big(W(2)\big) + \text{Var}\big(W(3)\big) + 2\text{Cov}\big(W(2), W(3)\big) \\ &= 2 + 3 + 2 \cdot 2 \\ &= 9. \end{aligned}$$

Thus, $X \sim N(0, 9)$. We conclude

$$\begin{aligned} P(1 < X < 2) &= \Phi\left(\frac{2-0}{3}\right) - \Phi\left(\frac{1-0}{3}\right) \\ &= \Phi\left(\frac{2}{3}\right) - \Phi\left(\frac{1}{3}\right) \\ &\approx 0.12 \end{aligned}$$

(c) Remember that if $0 \leq s < t$, then

$$W(s)|W(t) = a \ \sim \ N\left(\frac{s}{t}a, s\left(1 - \frac{s}{t}\right)\right).$$

(This has been shown in the Solved Problems Section of the Brownian motion section). We conclude

$$W(1)|W(2) = 1 \ \sim \ N\left(\frac{1}{2}, \frac{1}{2}\right).$$

Thus,

$$\begin{aligned} P(W(1) > 2|W(2) = 1) &= 1 - \Phi\left(\frac{2 - \frac{1}{2}}{1/\sqrt{2}}\right) \\ &\approx 0.017 \end{aligned}$$

31. (Brownian Bridge) Let $W(t)$ be a standard Brownian motion. Define

$$X(t) = W(t) - tW(1), \qquad \text{for all } t \in [0, \infty).$$

Note that $X(0) = X(1) = 0$. Find $\text{Cov}(X(s), X(t))$, for $0 \leq s \leq t \leq 1$.

Solution: We have

$$\begin{aligned} \text{Cov}(X(s), X(t)) &= \text{Cov}(W(s) - sW(1), W(t) - tW(1)) \\ &= \text{Cov}(W(s), W(t)) - t\text{Cov}(W(s), W(1)) \\ &- s\text{Cov}(W(1), W(t)) + st\text{Cov}(W(1), W(1)) \\ &= s - ts - st + st \\ &= s - st. \end{aligned}$$

33. (Hitting Times for Brownian Motion) Let $W(t)$ be a standard Brownian motion. Let $a > 0$. Define T_a be the first time that $W(t) = a$. That is

$$T_a = \min\{t : W(t) = a\}.$$

(a) Show that for any $t \geq 0$, we have

$$P(W(t) \geq a) = P(W(t) \geq a | T_a \leq t) P(T_a \leq t).$$

(b) Using Part (a), show that

$$P(T_a \leq t) = 2\left[1 - \Phi\left(\frac{a}{\sqrt{t}}\right)\right].$$

(c) Using Part (b), show that the PDF of T_a is given by

$$f_{T_a}(t) = \frac{a}{t\sqrt{2\pi t}} \exp\left\{-\frac{a^2}{2t}\right\}.$$

Note: By symmetry of Brownian motion, we conclude that for any $a \neq 0$, we have

$$f_{T_a}(t) = \frac{|a|}{t\sqrt{2\pi t}} \exp\left\{-\frac{a^2}{2t}\right\}.$$

Solution:

(a) Using the law of total probability, we obtain

$$\begin{aligned} P(W(t) \geq a) = P(W(t) \geq a|T_a > t)P(T_a > t) + \\ P(W(t) \geq a|T_a \leq t)P(T_a \leq t). \end{aligned}$$

However, since $P(W(t) \geq a|T_a > t) = 0$, we conclude

$$P(W(t) \geq a) = P(W(t) \geq a|T_a \leq t)P(T_a \leq t).$$

(b) Note that given $T_a \leq t$, $W(t)$ is normal with mean a. Thus

$$P(W(t) \geq a|T_a > t) = \frac{1}{2}.$$

Thus,

$$P(W(t) \geq a) = \frac{P(T_a \leq t)}{2}.$$

We conclude

$$\begin{aligned} P(T_a \leq t) &= 2P(W(t) \geq a) \\ &= 2\left[1 - \Phi\left(\frac{a}{\sqrt{t}}\right)\right]. \end{aligned}$$

(c) We can find the PDF of T_a by differentiating $P(T_a \leq t)$. We have

$$\begin{aligned}
f_{T_a}(t) &= \frac{d}{dt}P(T_a \leq t) \\
&= 2\frac{d}{dt}\left[1 - \Phi\left(\frac{a}{\sqrt{t}}\right)\right] \\
&= -2\frac{d}{dt}\Phi\left(\frac{a}{\sqrt{t}}\right) \\
&= \frac{a}{t\sqrt{2\pi t}}\exp\left\{-\frac{a^2}{2t}\right\}.
\end{aligned}$$

Chapter 12

Introduction to Simulation Using MATLAB (Online)

Chapter 13

Introduction to Simulation Using R (Online)

Chapter 14

Recursive Methods

1. Solve the following recurrence equations. That is, find a closed form formula for a_n.

   1. $a_n = 2a_{n-1} - \frac{3}{4}a_{n-2}$, with $a_0 = 0, a_1 = -1$.
   2. $a_n = 4a_{n-1} - 4a_{n-2}$, with $a_0 = 2, a_1 = 6$.

Solution:

(a) Characteristic equation:

$$x^2 - 2x + \frac{3}{4} = 0$$

By solving the equation, we get:

$$x_1 = \frac{1}{2}$$
$$x_2 = \frac{3}{2}$$

We define:

$$a_n = A(\frac{1}{2})^n + B(\frac{3}{2})^n$$

$$a_0 = 0 \longrightarrow 0 = A + B$$

$$a_1 = -1 \longrightarrow -1 = \frac{A}{2} + \frac{3B}{2}$$

By solving the equations, we get:

$$A = 1$$
$$B = -1$$

By substituting the values of A and B to the equation $a_n = A(\frac{1}{2})^n + B(\frac{3}{2})^n$, we get:

$$a_n = (\frac{1}{2})^n - (\frac{3}{2})^n$$

(b) Characteristic equation:

$$x^2 - 4x + 4 = 0$$

By solving the equation, we get:

$$x_1 = x_2 = 2$$

We define:

$$a_n = A2^n + Bn2^n$$

$$a_0 = 2 \longrightarrow 2 = A$$
$$a_1 = 6 \longrightarrow 6 = 2 \times A + 2 \times B$$

By solving the equations, we get:

$$A = 2$$
$$B = 1$$

By substituting the values of A and B to the equation $a_n = A2^n + Bn2^n$, we get:

$$a_n = 2^{n+1} + n2^n$$

3. You toss a biased coin repeatedly. If $P(H) = p$, what is the probability that two consecutive H's are observed before we observe two consecutive T's? For example, this event happens if the observed sequence is $THT\underline{HH}THTT\cdots$.

Solution:

Let A be the event that two consecutive H's are observed before we observe two consecutive T's. Conditioning on the first coin toss:

$$\begin{aligned} P(A) &= P(A|H)P(H) + P(A|T)P(T) \\ &= pP(A|H) + (1-p)P(A|T) \end{aligned}$$

$$\begin{aligned} P(A|H) &= P(A|HH)P(H) + P(A|HT)P(T) \\ &= 1P(H) + P(A|T)P(T) \\ &= p + (1-p)P(A|T) \end{aligned}$$

So:

$$P(A|H) = p + (1-p)P(A|T)$$

$$\begin{aligned} P(A|T) &= P(A|TH)P(H) + P(A|TT)P(T) \\ &= pP(A|H) + 0P(T) \\ &= pP(A|H) \end{aligned}$$

So, by combining the two results, $P(A|T) = pP(A|H)$ and $P(A|H) = p + (1-p)P(A|T)$:

$$P(A|H) = p + (1-p)pP(A|H)$$

So:

$$P(A|H) = \frac{p}{1-p(1-p)}$$

Thus, we obtain

$$\begin{aligned} P(A) &= pP(A|H) + (1-p)P(A|T) \\ &= pP(A|H) + (1-p)pP(A|H) \\ &= p(2-p)P(A|H) \\ &= \frac{p^2(2-p)}{1-p(1-p)}. \end{aligned}$$

Made in the USA
Las Vegas, NV
16 January 2023